THE LAST PANDA

George B. Schaller

With a New Afterword

The University of Chicago Press

Chicago and London

The University of Chicago Press, Chicago 60637
The University of Chicago Press, Ltd., London
 Published 1993
Paperback edition 1994
Printed in the United States of America

02 01 00 99 98 97 96 95 94 1 2 3 4 5 6

ISBN: 0-226-73629-6

Library of Congress Cataloging-in-Publication Data

Schaller, George B.
The last panda / George B. Schaller ; with a new afterword.
p. cm.
Includes bibliographical references (p.) and indexes.

1. Giant panda. 2. Endangered species. 3. Wildlife conservation.
I. Title.
[QL737.C214S28 1994]
599.74′443—dc20 94–27371
CIP

♾ The paper used in this publication meets the minimum requirements of the American National Standard for Information Sciences—Permanence of Paper for Printed Library Materials, ANSI Z39.48-1984.

To the memory of

Sir Peter Scott

George Schaller

Contents

"From one year's end to another, one hears the hatchet and the axe cutting the most beautiful trees. The destruction of these primitive forests, of which there are only fragments in all of China, progresses with unfortunate speed. They will never be replaced. With the great trees will disappear a multitude of shrubs and other plants which cannot survive except in their shade; also all the animals, small and large, which need the forest in order to live and perpetuate their species. . . .

"This Cosmos, so marvelous to those with eyes to see, is reduced to dullness by a blind and egotistical preoccupation with material things. Soon the horse and pig on the one hand, and wheat and potatoes on the other, will replace all these hundreds, these thousands of creatures—animal and vegetable—that God has created to live among us. They have the right to life and we annihilate them and brutally make existence impossible for them. . . . It is unbelievable that the Creator could have placed so many diverse organisms on the earth, each one so admirable in its sphere, so perfect in its role, only to permit man, his masterpiece, to destroy them forever."

Père Armand David
1875

大熊猫

Preface

Years have passed since I last saw giant pandas in the wild, yet their powerful image continues to impinge on my life. Pandas are creatures so gentle and self-contained that they still affect me by the force of their uniqueness, by their aura of mystery. Our research on pandas gained only superficial insights into their alien existence, yet to see a panda again in a zoo or photograph releases a cascade of memories, of snowbound forests, of tracking animals over mountains lost in fog, of a female panda named Zhen-Zhen methodically crunching shoots in the bamboo's shadows. Feelings cannot, however, be fully recreated, and experiences cannot be truly shared. Shorn of emotion, my years with the pandas can be summarized rather briefly.

On 15 May 1980, I walked for the first time through a forest inhabited by giant pandas. Although I did not see pandas that day, I did find their droppings and examined sites where they had fed on bamboo. Four and a half years later, on 10 January 1985, I spent my last day in their home. During this period I devoted the first two years to research in the Wolong Natural Reserve, at seven hundred eighty-five square miles the largest of China's reserves containing pandas. There, at a research camp named Wuyipeng, Fifty-one Steps, my wife, Kay, and I, with our Chinese colleagues, spent many months between December 1980 and June 1982. This period was followed by eight months in the United States, at the New York Zoological Society, which has been my employer for over a quarter century, where I worked on a scientific panda book. Early in 1983 I returned to China, both to continue research in Wolong and to survey other panda areas in Sichuan, the province where most pandas occur and where the project was based. This survey was useful in that it revealed the different ecological conditions under which pandas subsist, and it helped us select a second research site, the Tangjiahe Natural Reserve. I was mainly in Tangjiahe during 1984, studying pandas as well as Asiatic black bear. I wished to compare the ecology and behavior of these two species that are so similar in size and shape, and also to ascertain if Tangjiahe pandas differed in any significant way from those in Wolong.

After devoting more than four years to pandas, I left the project to my colleagues. Kenneth Johnson, a bear specialist from the University of Tennessee, had joined the project in 1984, as had Alan Taylor, then a botanist at the University of Colorado. Ken monitored pandas in Wolong while I concentrated on Tangjiahe, and Alan studied bamboo. Donald Reid came to China from the University of Calgary in 1985 to continue research in the two reserves while Ken participated in a panda census to determine the actual status of the species.

The function of foreign experts, as we were called, was to help initiate work and introduce our Chinese coworkers to new technology, ideas, and techniques, not to do routine research. I had fulfilled my obligations. We had learned much about the panda's intricate adaptation to bamboo by noting in detail what parts of this plant the panda ate at different seasons and by analyzing the nutritional content of bamboo in the laboratory. We had studied pandas' movements and daily activity cycles by radio telemetry, a method that allowed us to track individual animals by monitoring radio signals from a transmitter attached to a collar fastened around the animal's neck. We also defined the panda's conservation problems and the steps needed to alleviate them. A baseline of information was available, the tasks ahead were clear; others would fill in the gaps and details and implement our recommendations of measures needed to protect the species. And I must admit that my enthusiasm for a study wanes once basic insights have been gained. My mind becomes directed toward new and to me more urgent goals, in this instance a study of the unique wildlife of the Tibetan Plateau.

The panda project continued, carried on by an occasional foreigner and several Chinese, many of whom I did not know personally. I saw reports, read news in the press, and received word from colleagues. But I purposely did not return to my old haunts. I could easily have done so, for after leaving the pandas I continued to work in China studying Tibetan antelope, wild yak, snow leopard, and other high-altitude species whose future is threatened. I felt a need to distance myself from pandas in order to reduce my experience with both pandas and people to coherence. As a foreigner, often the only foreigner, isolated for months with Chinese coworkers, I found it difficult to achieve a true perspective on the project. I naturally have two images of China as well as of the panda, the one I brought with me and the one I saw there. My perceptions may not always be reliable; I know they are limited, and at any rate I find myself constantly amending them, truth sometimes shifting with my feelings. In this book I hope to offer the Chinese and the pandas an understanding based on shared sensibilities, not an easy task because both the Chinese and the pandas embody a blend of deceptive stoicism and warmth.

The panda project was the most difficult of all my studies, not because of any inherent problems in the research, which was easy in conception and execution, but because it was conducted under the auspices of two bureaucratic institutions: the Chinese government, with its vast hierarchical structure from the highest level in Beijing to the lowest in our camp; and World Wildlife Fund International (WWF, now World Wide Fund for Nature) far removed in Switzerland from the pandas, yet determined to control the local situation. The two institutions understood the enterprise imperfectly and misunderstood each other frequently.

I had spent years in various countries—Pakistan, India, Tanzania, Zaire, Brazil—studying wildlife, and had dealt with officials at varying levels of government. A biologist must not only study nature but also induce action on behalf of conservation, and this guarantees problems for any project. Action can be initiated only at a political level, conservation being more a social matter than a scientific or technological one. I was accustomed to problems, sometimes caused by the absence of a shared goal and vision, at other times by muddled communication. But nothing could have prepared me for the fundamental barrier to comprehension that existed between the aims and methods of the World Wildlife side and the Chinese side, as each referred to the other. On a personal level, I had never been where relations were so cordial yet inarticulate, where my freedom to do even the simplest task or veer even slightly from a rigidly prescribed course was so restricted, where all my actions were so unrelentingly scrutinized and reported, and where my presence was treated with such wariness.

Of course part of the problem was that the Chinese were just emerging from the ten-year chaos of the Cultural Revolution that began in 1966, a period in which Chinese society, strained somehow beyond endurance, moved from self-criticism to self-destruction. Anything foreign was reviled and anyone with foreign contacts was labeled a reactionary, harassed and imprisoned. The "open door" policy began in 1978, and thereafter social changes occurred at a momentous pace. The China of 1980, when I arrived, was vastly different from the China of 1991, the year when this chronicle ends. Above all, there was by the mid-1980s a new atmosphere of hope. But the Chinese character, shaped by five thousand years of civilization, naturally had remained the same throughout: the philosophic objective to preserve dignity, the orientation to the past, the restraint in social relations, especially with foreigners. However, as people were given new freedoms, as "the mental shackles imposed by years of leftist ideas" were slowly removed, and as the objects of their modest cravings became more attainable, the Chinese with whom I had contact grew less watchful and treated me, the *weiguoren* or foreign-country person who had been thrust into their midst, with extraordinary kindness.

Some of this hope and freedom died on 4 June 1989. On that day, in the early morning hours, the government used the People's Liberation Army to kill hundreds of students and workers who had gathered in Beijing's Tiananmen Square for an exuberant protest movement against corruption and for human rights and free speech. In a menacing reminder of the Cultural Revolution, the government afterward investigated, purged, arrested, and even executed those with "bourgeois liberal inclinations" in a hopeless effort to reassert ideological control. For seven weeks in April and May of 1989, at the height of this movement, an exhilaration had swept through Beijing, Shanghai, Xian, Chengdu, and other cities, but the harsh reprisals intimidated the public. When I returned to China in August of that year, there was a veneer of normal city life. But once again, as a decade earlier, people hesitated to talk freely with foreigners; they were dispirited and burdened with a smoldering malaise.

The panda project was initiated just after a turbulent fragment of time in China's history. I felt privileged to be in China during this period—the only decade of peace the country had seen this century. It gave me the opportunity to study an animal that is considered a national treasure and to live and travel in remote mountains where often no foreigner had been for half a century. I came to admire my coworkers for their sense of humor, tact, and acceptance of difficult circumstances. And as I learned to understand them better, I viewed with sympathy, though not always with passive acceptance, what to my Western mind seemed to be insensitivity, arrogance, intrigue, and entitlement mentality. My memories dispose me to filter out the dark side of the project, aspects that sometimes left me terribly dejected. Why are memories so indulgent, so idealized by nostalgia? Also, I liked working in China with its unique research opportunities and urgent conservation issues, so much so that I spent over three times as many years there on wildlife studies as in any other country.

Yet I do not want to sound overly tolerant or too sympathetic; my admiration for China need not make me an advocate or apologist. Friendship is not blind. And my admiration for WWF as an institution does not diminish my exasperation with its penchant for highhandedness in dealing with the project. It would be easier and more enjoyable to write an inspirational account of selfless dedication and uplifting ideals, of natural grandeur and compelling truths, and finally, with a shout of triumph, proclaim the panda saved, a noble venture concluded. Such had been my aspirations for the project, but the final disillusioning reality is otherwise.

To convey what happened with fidelity, I have to discuss panda politics as much as the pandas themselves. A conservation project is always divided between politics and science, and any book about such a project ought to reflect the constant interplay between the two. Yet accounts of battles to save whales, rhinoceroses, and other species whose fate are of public concern tend to shy away from disclosing the true conservation conflicts, the basic issues of human greed and indifference. The panda has become a lucrative commodity, in its own way little different from the elephant, and thus has revealed ignoble traits hidden in some individuals and organizations. Indeed, political problems overshadowed and constantly intruded upon both the scientific and conservation efforts of the panda project. To hide this aspect of the work would be a grave disservice to the panda. I realize that no account can bear too much truth, and this made me reluctant to write at all. Criticism is irritating, not well received, and often misunderstood. Those readers of this book who have traveled in China only as tourists or on brief assignments may have left the country with impressions far different from mine. However, those who have lived and worked in China for a year or more, as teachers, journalists, or in other professions, may have empathy with the difficulties experienced by the panda project. But whatever a reader's background, I hope to arouse every emotion except indifference to the plight of the panda.

I am ambivalent as well about adding yet another book on endangered species to the ever-increasing library on that subject, especially about an animal as well documented as the panda. In what other species are all captives known by name, Mei-Mei, Li-Li (double names are a sign of affection)? What other species has its every birth, illness, reluctant romance, and death announced by the news media? For anyone who wants to read about pandas there are already many articles and several books. The two most informative popular books are *Men and Pandas,* by Ramona and Desmond Morris, published in 1966, a good account especially about the Western expeditions that went to kill or capture animals, and *Pandas,* by Chris Catton, published in 1990. In addition, several recent popular books have appealing photographs of captive pandas, often in a natural setting. The text in these books ranges from trivial (*The Secret World of Pandas,* edited by Byron Preiss and Guo Xueyue) to competent (*The Bamboo Bear,* by Clive Roots). The *Giant Pandas of Wolong,* by my Chinese coworkers and myself, published in 1985, is a scientific treatise describing in detail what this book covers only in general terms.

The many books on dead or dying species almost seem to lessen the intensity with which one views extinction. Repetition begins to trivialize a terrible event. No matter how much we may decry species extinction, nature still remains peripheral to the consciousness of most people. Are these books mostly bought as tokens of a belief that nature does count? Too often treatises on endangered species seem to be mere memorials, with all the finality that this implies, accounts of those animals whose drama has irrevocably ended: the dodo, passenger pigeon, great auk, and Steller's sea cow. My own emotions at least cannot respond to the ever-expanding list of dying animals; my capacity for concern is finite.

Yet there is a justification for these volumes. Telling and retelling is a moral imperative; forgetting is a luxury we cannot afford. Neglect is a form of abuse. There are those who think that if only we can hold onto our biological diversity, all those millions of species of plants and animals, into the next century, destruction will end—yet there is no indication that it will. So we must at least record our experiences with the hope that our writings will encourage action to preserve species and stimulate a unity of compassion. Even in a truly moral world, destruction would not end, but at least we would view nature with finer sentiments, based on a revolution in the spirit of humankind. We would adjust our values and priorities and develop a land ethic that decries waste and needless destruction. Such changes cannot come through passion and strident rhetoric but only through a new concept of ourselves, a new design in the strategy of human survival.

Perhaps the panda in a small way can help change our concepts. Its seeming simplicity permits us to discern the qualities that make it so alluring. Having transcended its mountain home to become a citizen of the world, the panda is a symbolic creature that represents our efforts to protect the environment. Though dumpy and bearlike, it has been patterned with such creative flourish, such artistic perfection, that it almost seems to have evolved for this higher purpose. A round, rather flat face, large black eye patches, and a cuddly and clumsy appearance give the panda an innocent, childlike quality that evokes universal empathy, a desire to hug and protect. And it is rare. Survivors are somehow more poignant than casualties. Together, these and other traits have created a species in which legend and reality merge, a mythic creature in the act of life.

We are fortunate that the panda is still with us, that our evolutionary paths have crossed. Values should not, I suppose, be assigned to creatures to whom values are unknown, and who all have equal claim to our fascination and respect. Still, the panda would seem more of a loss than a primrose or piranha, for it epitomizes the stoic defiance of fate and stirs our emotions with pity and admiration. If we lose the panda, we will never look on its black-and-white face again, its evolution will stop, its unique genetic code destroyed; its name will soon have little more significance than that of thousands of other species listed in the dusty catalogues of the world's museums, *Ailuropoda melanoleuca,* meaning the "black-and-white panda-foot." As the obscurity of centuries separates the animal from us, we will be left with only mementos, a few massive bones, some faded hides. The pandas' lives will be forgotten, gone from our collective memories like the moa and mammoth. What a melancholy fate for so extraordinary an animal. All creatures, of course, are transitory, flourishing for a while, then fading away. However, pandas are survivors who were present several million years ago, before humankind became human, and they have outlived many other large mammals that vanished during the upheavals of the Ice Age. Their time as a species should not yet be over.

The 1980s were among the most momentous and the best-documented years in the panda's long evolutionary history. Unless they are chronicled, the events of these years will not endure. I went to China in the role of scientific emissary to understand and record, to become an interpreter of an inarticulate life, and to leave a legacy of the panda's existence. This book is a small part of that legacy.

The dedication and spirit of cooperation of several organizations and many individuals made the panda project possible. Most of those who helped are mentioned in this book, and I would like to express my gratitude to all collectively. In addition, there are some whose contribution was so sustained, selfless, and important that I owe them a special debt.

I would like to express my appreciation to the World Wide Fund for Nature (WWF) for asking me to study the panda and for entrusting me with this exceptional project. Mark Halle and Christopher Elliott helped guide the program and provided much assistance.

The New York Zoological Society generously permitted me to neglect my job as director of its conservation division, Wildlife Conservation International, so that I could devote myself to pandas. Although the project was officially financed through an agreement between WWF and the Chinese government, the New York Zoological Society also provided major financial support. It assumed most of my expenses, it paid for field equipment and supplies and for laboratory analyses, it provided the services of several staff members, and it gave a one-year stipend to Donald Reid, a total donation of nearly half a million dollars. For this financial and moral support I am especially grateful to William Conway, the general director of the New York Zoological Society. He also came to China to assist with the design of the Wolong breeding station. Two of the society's veterinarians, the late Emil Dolensek and Janet Stover, each made two trips to help with the breeding effort and management of captives, interruptions in their busy schedules that I deeply appreciated.

Michael Pelton of the Department of Forestry, Wildlife, and Fisheries, University of Tennessee, not only made the computer and library facilities of his department available but also provided the project with three assistants. Howard Quigley worked with me to place the first radio collars on pandas, Patrick Carr analyzed radio telemetry data by computer, and Kenneth G. Johnson became a major researcher in the panda project, as described in this book.

Alan Taylor, currently at Pennsylvania State University, Donald Reid, now at the University of British Columbia, and Kenneth Johnson at the University of Tennessee conducted their panda research with admirable skill and dedication. The list of publications at the end of this volume attests to their important contribution to panda ecology. And after I left the project, all kept me informed about the behavior of pandas and people at Wolong and Tangjiahe, a courtesy for which I am most grateful.

Ellen Dierenfeld, then at Cornell University and now the nutritionist at the New York Zoological Society, first contacted me in 1980 to describe her research on panda nutrition at the National Zoo in Washington, D.C., and to offer her assistance to the project, especially with chemical analyses of bamboo. Hundreds of bamboo samples were subsequently analyzed by Ellen and by James Robertson in the laboratory of the Department of Animal Science at Cornell University. Their contribution was critical to an understanding of the panda's life-style. Ellen also patiently gave me a short course on animal nutrition, a subject about which I was embarrassingly ignorant. Her insights helped to elucidate the energetics of the species. I am greatly indebted to her.

Nancy Nash, an American journalist based in Hong Kong, was instrumental in establishing the panda project, and she contributed to it in innumerable ways over the years. She continually gave valuable advice based on her sensitivity to Chinese perceptions, hosted foreign and Chinese travelers associated with the project in Hong Kong, shipped urgently needed equipment, and acted as liaison between Wolong researchers and Beijing, to name just a few of her contributions. Her many gestures of kindness, such as sending magazines and delicacies to our isolated camp, were always appreciated. So effectively and with such devotion did she promote panda conservation that the Chinese have referred to her as Miss Panda.

My contact with Wang Menghu of the Ministry of Forestry began in 1980. First as coordinator of the panda project and then of some of my subsequent work, he always showed his commitment to conservation. Every problem, trivial or major, came to him, and he did his utmost to satisfy and balance the demands of foreigners and those of the various factions among the Chinese. I greatly admired his tenacity and skill in overcoming obstacles. He was a *lao pengyou,* old friend, in the true sense of this Chinese phrase.

Hu Jinchu, a biology professor at Nanchong Teacher's College in Sichuan, was my principal Chinese counterpart, and I look back upon our close association with pleasure. A fine naturalist, Hu Jinchu taught me much about the birds and other forest creatures. It was a delight to be in the field with him, for no matter how depressing the weather or how rough the terrain, he enjoyed being there. He retained a sense of adventure and a heart responsive to the natural world.

Qiu Mingjiang was one of several interpreters who assisted the project over the years. Bright and fluent in English, he came to us with the dreams of youth, his enthusiasm, energy, and idealism undiminished. He became much more than an interpreter as, eager to learn, he joined in the field work, translated two panda books from English into Chinese, and became so imbued with a research and conservation ethic that he not only worked tirelessly on behalf of the project but also returned to college to broaden his biological knowledge.

Others, too, deserve a special expression of gratitude. Iain Orr sent a steady stream of articles to keep me informed about China's environment. Millicent Se Yung and Fay Loo assisted the project as interpreters and gave much good advice. Douglas Kreeger donated winter boots, sleeping bags, and other equipment. Lynn and the late Irene Saunders received me hospitably in their Beijing home, as did the Austrian ambassador, Wolfgang Wolte, and his wife, Ursula. Donald Bruning and Andrew Laurie supplied me with much information about pandas.

With eloquent brush, Hank Tusinski created the paintings on pages v, vi, viii, and 278.

I am indebted to William Conway for reading the manuscript critically. Susan Abrams, my editor at the University of Chicago Press, greatly improved the manuscript with her editorial skills and insights.

Above all, I wish to thank my wife, Kay. As in my other projects, she participated in every aspect of the work to such an extent that the results are almost as much hers as mine. She spent long nights awake in a tent to keep track of radio collared pandas, she analyzed the contents of panda droppings, and she edited and typed all reports, articles, and books, including this one. As always, she made a home for us in the wilderness, a place of human warmth in a bitter environment. She was someone with whom I could discuss concerns, she was a buffer for disappointments. When at times we had to be apart for several months, my joy in the work diminished and I knew then how much she is the focus of my life.

Prologue

The ridge lunged upward like a dragon's spine bristling with fir and birch, and clouds were low and flying out from the mountains. Snow from a late-winter storm balanced on boughs and logs. When a riffle of wind stirred the branches, the snow drifted down in crystal veils that added a ghostlike radiance to the forest. Bamboo grew in the understory, the crowded ranks of stems claiming the hillside so completely that the light beneath the bamboo's canopy was a translucent undersea green. The sunless scent of moss and moldering wood choked the gloom. The bamboo was rigid with frost, and a dense silence hung over the ridge; there was no movement and seemingly no life.

In the stillness, leaves suddenly rustled and a stem cracked like breaking glass. Shrouded in bamboo was a giant panda, a female, slumped softly in the snow, her back propped against a shrub. Leaning to one side, she reached out and hooked a bamboo stem with the ivory claws of a forepaw, bent in the stem, and with a fluid movement bit it off near the base. Stem firmly grasped, she sniffed it to verify that it was indeed palatable, and then ate it end-first like a stalk of celery. While her powerful molars sectioned and crushed the stem, she glanced around for another, her movements placid and skillful, a perfect ecological integration between panda and bamboo. She ate within a circle of three feet, moved a few steps and ate some more, consuming only coarse stems and discarding the leafy tops; she then sat hunched, forepaws in her lap, drowsy and content. Within a circle of three thousand feet was her universe, all that she needed: bamboo, a mate, a snug tree-den in which to bear young.

Minutes later, she ambled in her rolling sailor's gait to a nearby spur where among gnarled rhododendrons she halted. No bamboo grew here. A shaft of sun escaped through a fissure of cloud and penetrated the twilight. Among bamboo the panda's form and color had seemed blurred and difficult to define; now in sun the panda shone with sparkling clarity. Near her was a massive fir. She knew that tree: it was a landmark, it defined the edge of her favored haunts, it served as a scentpost. The tree's many dimensions helped give her an identity. The snow around the tree was unmarked by tracks, but when she sniffed the bark, she learned that a male had marked the site with his anal glands a few days before. Though she fixed the scent in her mind, she did not cover his odors with hers.

She angled down to the nearest bamboo patch and there once again foraged, the recycling of bamboo being the essence of her existence. She lived leisurely. Alone in these heights, the panda conveyed a sense of absolute solitude, an isolation that was almost mythic. A flock of tit-babblers skittered like airborne mice through the bamboo above her head, yet her small dark eyes showed no awareness. Having eaten, she rolled over to sleep, her body at rest in the snow against a log, her dense coat making her impervious to the elements.

From below, near where forest gave way to field, came the sound of an ax. The bamboo around her like armor against intruders, she listened and then moved away, shunning any possible confrontation. She traveled on a private path along the slope, insinuating herself from thicket to thicket, moving like a cloud shadow, navigating with precision through the sea of stems, with only her tracks a record of her silent passing.

1

"Every Journey Begins with the First Step"

May and June 1980

The path climbed up the mountainside toward a spur, above the spring fields of potato and maize, until the Pitiao River was just a murmur and all around peaks rose into a slate sky. The edge of the path was heavy with tangled growth. Now and then we loitered to admire an anemone or wood sorrel or other small flower, to take note of a lavender rhododendron aflame, to examine bamboo shoots, thick as a thumb, piercing the shade. Spiny seed capsules of last year's hazelnuts littered the ground like swarms of tiny hedgehogs. Above, among birch and fir, a Himalayan cuckoo called, a mellow boo-boo-boo.

"Look there. That's the Wolong dragon," exclaimed Liu Yanying, our interpreter, pointing up-valley. Xiao Liu, as we called her, *xiao* being an affectionate diminutive meaning "small," was our link to the twenty-one Chinese who curved up the path behind us. The dragon's blunt head rested at the base of a slope among village huts and its tail trailed upward into cloud. Mountains are full of dragons—serpentine rivers, smoke drifting from squat huts, sinuous crests—but this ridge looked indeed like a huge creature frozen in time. Once long ago, the story is told, a dragon flew over these mountains and was so enraptured by their beauty that it decided to rest a while. It was here still, this resting dragon *Wo-long,* for whom the reserve was named. Dragons are divine creatures, symbols of luck, of life; Wolong seemed an auspicious place.

On top of the spur was a small level spot the size of a room. Nailed to a rhododendron was a sign in Chinese proclaiming "Path of Welcome to Guests." We gathered there, in the shadow of pines and bamboo, breathing deeply after the stiff climb, the air cool with the chill of spring. Security police, officials, journalists, biologists, a doctor, and others, all had come to escort four foreigners. We were at the forest's edge, the first Westerners in decades to enter the giant panda's realm. And, more important, we were the first ever to be invited there by the Chinese government. The panda expeditions of the 1920s and 1930s had simply taken advantage of the political turmoil, wars, and corruption of that period to kill and capture animals. The leader of our small delegation was Sir Peter Scott, writer, artist, chairman of WWF, a giant in the conservation movement of the twentieth century, who died in 1989. Sir Peter's presence was particularly appropriate, for twenty years earlier he had selected the panda as the symbol of WWF and designed that organization's distinctive panda logo. His wife, Lady Philippa Scott, accompanied him. And Nancy Nash, the Hong Kong–based journalist whose enthusiastic devotion to conservation resulted in the first contact between China and WWF, was here too.

The path now traced the forested contours of a slope as we entered the Choushuigou (*gou* means valley), a small drainage girdled by high ridges. With relief I noted that the terrain, though rugged, could be traversed without difficulty. Yesterday, when we had entered Wolong, the road had wound for miles along a river gorge, turquoise waters storming among boulders, veils of cloud clinging to rock faces. Patches of forest huddled either low on the flanks or perched on the crests: vertical mountains and water, an idealized Chinese landscape. It had looked spectacular and intimidating, and almost impossible for field work. We traveled now in single file, our footsteps muffled on the soft earth, our voices subdued as if in a holy place. Two pale-green objects lay at the edge of the path, each spindle-shaped and about six inches long and two inches wide. Panda droppings! I knelt and cupped one in my hand. It consisted of undigested pieces of bamboo stem all neatly aligned and held together with mucus; it smelled sweet, like freshly cut grass. Carefully I passed the fragile treasure to Sir Peter, as all the others crowded near to discover the cause for our reverence. Sir Peter and I grinned at each other, tremendously pleased and oblivious to what our hosts might think about our delight in fondling feces. Some days are marked for recollection. "A grand day, isn't it," commented Sir Peter with British understatement.

Now Professor Hu Jinchu, China's leading panda expert, became our guide. He had studied pandas since the mid-1970s and two years previously had established the Wuyipeng research base in the Choushuigou. Slightly older than I, he was of medium build with a round face and short hair that leaped straight up from his head. We had been quietly evaluating each other, our spirit, our commitment, knowing that we might soon share responsibility for the panda research. But I had discerned little beyond his being gentle-mannered and that he laughed frequently from nervousness. He pointed to a pine where a panda had clawed and bitten off a slab of bark, leaving an oozing wound. Further on, in a tiny clearing between clumps of bamboo, he showed us a pile of hairy husks. A panda had sat there and stripped bamboo shoots before eating the succulent centers; nearby were soggy, pale droppings, the fibrous remains of shoots. Downhill to the left, a trail of broken herbs revealed an animal's route. "*Daxiongmao*," said Hu Jinchu, using the Chinese name, large bear-cat, for the panda. With a laugh, he followed the trail, pushing arching bamboo stems aside with his arms, obviously delighted by our intense interest in examining the artifacts of a panda's passing. He soon found a site where a panda had bitten off two bamboo stems, taken a few mouthfuls, and discarded the leafy tops. Why would a panda ignore leaves in favor of tough woody stem, I wondered?

"Is this the best place to study pandas?" asked Wang Menghu as we examined the bamboo remains. Wang Menghu had told us that his name means "dreaming of tigers," and he felt that somehow the name had helped turn his destiny toward wildlife. He was in charge of China's nature reserve system for the Ministry of Forestry, and for panda conservation too. A small man in his forties, with smiling eyes, thinning hair, and a talent for manipulation, he expectantly awaited an answer.

"I don't know," I replied. Xiao Liu translated, although Wang Menghu obviously understood some English. "It's the only place we've visited so far. We need to survey several areas and then select the best study site."

Wang Menghu smiled enigmatically, and noted: "There is a proverb in our country, 'Every journey begins with the first step'."

Later, we retraced our route down the mountain, so euphoric we almost ran down the winding path. Crossing the Pitiao River on a hanging bridge that swayed underfoot, we returned to our cars waiting in a hamlet. The reserve headquarters was only a few miles away at a wide bend of the valley, a dreadful huddle of large cement buildings. These had housed several thousand loggers before the reserve was established in 1975. A floor of one building had been newly converted into guest rooms and one large meeting room.

All meeting rooms in China look alike: overstuffed chairs and sofas along the walls, and close at hand low tables with thermoses of hot water and covered porcelain cups from which to sip green tea. Officials spend an inordinate amount of time in meetings sharing their responsibilities; the habit is recognized by the government as an "unhealthy tendency" that cuts down on productivity. I had always managed to contain my enthusiasm for meetings of any kind. Yet since my arrival in China on 2 May, Nancy Nash and I—the Scotts did not join us until Chengdu—had already had a surfeit of meetings.

There had been a daylong meeting to discuss China's nature reserves. The first reserves, Wang Menghu told us, had been established in 1958 and the first panda reserves in 1963; over sixty reserves existed now, and there might be two hundred and fifty or one percent of China's land area, by 1985. Most panda reserves are under provincial control, but three of them, including Wolong, are under the central government. A total of 14.27 million yuan (about 9.5 million dollars) had so far been spent on the panda reserves, Wang Menghu stressed, and the figure would treble by 1985.

Another day we listened to Zhu Jing, a mammalogist from the Academy of Sciences, as he reviewed past panda research by Chinese. Pen Hongshou had made a habitat survey along the northern limit of the panda's range in 1940. A team of scientists censused pandas in Wanglang Reserve in 1968 and 1969 but the Cultural Revolution terminated work. At the end of April 1975 and in early May 1976 about three thousand people were sent into the mountains for three days to census pandas and delineate distribution. The results revealed that pandas survived only in six discrete mountain areas covering about twelve thousand square miles. The estimate of numbers was one hundred to two hundred in Shaanxi, about one hundred in Gansu, and eight hundred in Sichuan province, a total of some eleven hundred pandas.

There was a meeting at the Institute of Botany to provide background information on bamboo ecology. Later Wang Menghu gave us a briefing on last year's initial meeting with WWF. This was followed by two days of discussion on a Protocol for Establishment of a Research Centre for the Protection of the Giant Panda, as it was ponderously called. Interspersed with the meetings were tours to the Great Wall, Forbidden City, Beijing zoo, and other sights. All this was done leisurely, the time used to get acquainted, to establish rapport, to evaluate a potential social relationship that would, it was hoped, persist for years. That day in the Choushuigou, Wang Menghu remarked casually to me, "You have spent half your life elsewhere and now you will spend the other half in China."

Wolong's clouds were heavy and low, and rain beat on the tile roof the morning after our trip up the mountain. Another meeting was called, this

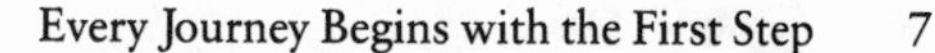

Distribution of giant pandas. The information is based on data collected by Ken Johnson and his colleagues during the 1985–88 panda census. The numbers on the map refer to reserves in which I did field work: (1) Jiuzhaigou, (2) Wanglang, (3) Tangjiahe, (4) Wolong, (5) Fengtongzhai, (6) Mabian.

one about the reserve. We were told that so far ninety-six mammal, two hundred and thirty bird, twenty reptile, and fourteen amphibian species had been recorded. Of these, the panda, golden monkey, white-lipped deer, and takin were in China's category I, species fully protected and under central government jurisdiction. Wolong's headquarters lies at an elevation of sixty-five hundred feet, average annual rainfall is forty inches, and the humidity averages eighty percent; there are only one hundred and eighty frost-free days. Local people were said to be kind to pandas: in October

1978, a sick panda had gone to a commune family where it was fed on sugar and rice porridge for three days until it recovered and returned to the forest.

This outpouring of facts in meeting after meeting was less to inform us than to emphasize that much research had already been done, and that, as Wang Menghu noted, WWF was needed mainly to establish the research center, as had been agreed the previous year. The implication was that to get a research center the Chinese would tolerate a foreign biologist or two. However, my private conclusions were different: little detailed field work had been done and a research center was *not* needed. I remembered when during our visit to Wuyipeng a few days before I had asked to see all panda research notes from the past month. I was interested in learning what kind of information the team recorded. After much scurrying, someone produced a single sheet of tissue-thin paper. "No, no," I said, "please show me a whole month's data, not just today's." There was an awkward silence: the single sheet represented a month's effort.

That afternoon we visited Hetauping, Walnut Terrace, about five miles down-valley from headquarters. It was the site proposed for the research center. There the river was pressed by mountain shadows, a forested slope on one side and a slanting cliff on the other. Why crowd a research center onto two small river terraces in a remote valley, cold and damp, when necessary chemical analyses could be done in the nearby city of Chengdu with its universities and laboratories? Why not convert an empty building at headquarters into a base with offices, library, herbarium, and so forth? How naive I was to assume that Western logic resembled Chinese logic.

Again it rained much of the night and into the morning. But as we climbed up a gorge into the Yingxionggou, a valley adjacent to the Choushuigou, a clear gray light filled the day. The trail, a route once used to extract timber, tunneled through rock at three places. Soon the gorge opened into a valley, and we reached what was termed a panda breeding facility, though no animals had as yet reproduced there. The pandas were housed in an open-sided shed with small, barred, and barren cells. Seven animals hunched there in cold winds that funneled from the slopes into the gorge. There was an outdoor enclosure with natural vegetation including bamboo, but the pandas were obviously not permitted to use it.

I was dismayed by the conditions. For days we had been told how precious the panda was, how proud China was of the animal. All through the 1970s, the government had used the panda as a political tool, giving treasured pairs to certain countries—Japan, the United States, England,

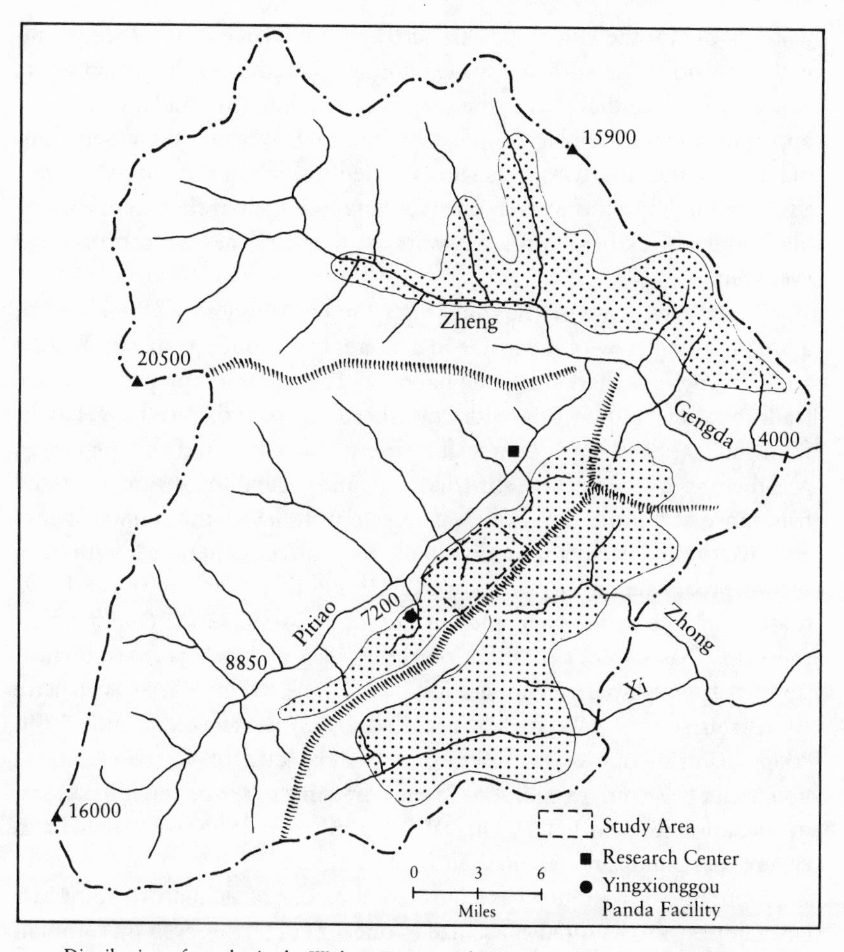

Distribution of pandas in the Wolong Reserve. The numbers indicate elevation in feet.

France, Spain, Mexico, North Korea—as a gesture of friendship. Zoos throughout the world vied for pandas; one American zoo even offered one thousand Holstein cows for a pair. One of my tasks, I felt, should be to clean up the place, make the pandas comfortable, introduce good management. I thought only of the pandas, unmindful of the fact that my Western habit of wishing to improve and alter things would clash with an enormous local inertia, based in part on the feeling that hard work leads nowhere.

My dark thoughts about the dismal facilities were dispelled, however, when the keepers brought out two panda young, seven to eight months old. They weighed about thirty pounds, still small enough to embrace if one watched out for their agate claws, sharp and curved like a cat's. Most mammals are unobtrusive; the panda cannot be ignored. And these youngsters had all the allure that makes the species so special. They had a vulnerable appearance, and the black patches around their wistful eyes added a humanizing detail that was irresistible. I tried to look at the animals objectively, telling myself that everyone would ignore them if they were entirely black like a black bear. Still, they were the most endearing creatures I had ever seen.

Next day we drove up the valley high into the mountains, where we had a breathtaking view of the inside of a cloud. It was our last day in Wolong and that night we had a farewell banquet. There was a long procession of traditional dishes, starting with cold sliced meats and spiced vegetables, followed by hot foods of wonderful variety, many with red chili pepper to give them spirit: bean curd, stir-fried pork and vegetables, shrimp, steamed fish, more and more until one was replete well before the soup at meal's end. We toasted each other individually, as courtesy demanded, with beer, or sweet red wine, or *bai jiu,* a powerful clear grain liquor like raw arak. We toasted and ate and toasted and ate and toasted some more. *Gan bei,* bottoms up. Toasts were ritualized, our hosts reciting stock phrases, invocations of friendship and cooperation: "Welcome to our guests who have come from so far," "We are new friends and will soon be old friends," "To cooperation in our work." Everyone was expected to say something. I noted that "Western expeditions in the past came to shoot and kill pandas, to take animals away from China. WWF is different. It has come to help, to ensure that pandas remain in their native home."

Attending the banquet was Chiu Xiaochiu, one of Sichuan's leading artists. Chinese artists often specialize in one art subject or even one animal. Chiu Xiaochiu, an elfin man with thick glasses that gave him an expression of tidy dreaminess, was partial to pandas. That afternoon he had spread a sheet of rice paper on a table and with a dozen or so sparse and deft brush strokes—legs, ears, eyes, curving back—had created a moment in a panda's life. The animal had been there at rest, satiated and slumped against bamboo. Chiu Xiaochiu had inscribed the painting to me and stamped it with his personal seal, then gently handed me this treasure. Each of us had received a painting. The panda in art as in life is such a bold, simple, and decorative form that it puzzled me why pandas were painted so seldom before the 1950s. No pandas adorn ancient scrolls or murals, yet Chinese art swarms with wildlife. The turtle represents wisdom, the crane long life, the bat good luck. The tiger is a symbol of belligerence and power, and the deer of riches, but the panda lacks such spiritual resonance; it has no mythic reality, no allegory, in Chinese history. Perhaps it was left out of this pantheon because of its self-effacing habit of hiding in bamboo. It is an animal of illusion, even more than a dragon.

It is traditional in China to commemorate special occasions with poems. But Chiu Xiaochiu did not at that banquet compose an ode to the panda. Instead he recited, "A dragon descended from the heavens, seeking an earthly home, a Palace of Nature, a Land of Happiness. . . ."

"He is a Buddhist," exclaimed Nancy in her expansive way.

"He is not a Buddhist! He is a Marxist!" corrected Xiao Liu emphatically, wrenching us back to the present.

WWF became involved in the panda project only through a fortuitous accident. The organization had hired Nancy Nash to work at its Swiss headquarters as a public-relations consultant for three months in mid-1979.

"Since you have the panda as a symbol," she asked one day, "why aren't you in touch with China about a panda study?"

"We've tried," she was told, "but it's impossible."

"Shall I try?"

"Yes," was the reply.

With typical initiative she prepared a proposal that suggested a meeting between a WWF delegation and relevant Chinese organizations to discuss how best to promote conservation jointly in the country. This proposal she channeled to Beijing through contacts with *Xinhua,* China's news agency, in Hong Kong. It was a timely move. China had announced its program of Four Modernizations—agriculture, industry, defense, and science—the previous year. Pragmatism had won out over the country's desire for seclusion. Wrapped in a pride based on centuries of glory, the Chinese had long avoided what they saw as the crude cultures surrounding them; but now they needed technology. The Chinese agreed to a WWF visit.

From 19 to 29 September 1979, a WWF delegation, including Sir Peter Scott, Lee Talbot, public-relations director David Mitchell, Nancy Nash, and Secretary General Charles de Haes, visited China. Many conservation topics were discussed. A China-WWF joint committee was formed with three members from each side to meet annually. A memorandum of understanding was signed, and it included an agreement to develop a project on "conservation of the Giant Panda, including establishment of a research centre on this species and systematic research on its biology." It was a historic meeting, but it also laid the groundwork for such serious misunderstandings that the panda project almost died before it was born. By signing the memorandum, WWF in effect agreed to a research center without exploring what this might imply. And the Chinese side designated three of its agencies—Ministry of Forestry, Chinese Academy of Sciences, and the Environmental Protection Office of the State Council (now the National Environmental Protection Agency)—to sponsor the project jointly. That decision lacked wisdom. Chinese count mainly on themselves and on the loyalty of their family, and to a much lesser extent on their workplace or unit. Beyond those bonds there is little sense of duty unless there is *guanxi,* a personal relationship. Consequently each unit is self-contained, competitive, and uncooperative except as dictated by its leader. *Po po tai dou,* as the Chinese say, too many mother-in-laws. The panda project had far too many.

Every pleasure comes with a shadow. For three weeks the Chinese had organized delightful banquets and excursions, hosting us with great formal warmth and solicitude. Now, back in Beijing in a room of the Beijing Hotel we met to begin serious negotiations for the project. Wang Menghu talked for several hours in great detail about his highest priority, the research center.

"The research laboratories will comprise eight hundred square meters and about twenty rooms. The cost of construction will be about two hundred and fifty yuan per square meter. . . . The living quarters will consist of sixteen hundred square meters. An associate professor will receive seventy square meters of living space, a technician thirty square meters. There will be about thirty scientists and technicians. . . . A ten-hectare outdoor enclosure will be built for twenty pandas. It will require five kilometers of fencing, three meters high; it will need twenty-five hundred poles at a cost of ten yuan per pole. . . . The research laboratories will be on one side of the river, the enclosure on the other. A bridge will cost three hundred thousand yuan. . . . A two hundred and fifty kilowatt hydroelectric station will be built."

On and on. We were stunned. Total construction costs would amount to two million dollars. The idea of a captive breeding facility had merit, for the deplorable one at Yingxionggou should be closed. But what did a big laboratory have to do with panda conservation? Besides, I felt the planning was backward: the Chinese had decided on amount of floor space and number of technicians without any discussion of the kind of research that would be done. I did not know then that in China the size of a building, no matter how inappropriate, makes a statement about a project's importance.

Sir Peter responded at length. "The main long-term objectives are to prevent further decline in the number of pandas as a result of habitat destruction, and to increase numbers through field research and captive breeding. The purpose of the latter is to release pandas into the wild and supply zoos, to eliminate the need to capture animals in the wild." He went on to discuss the importance of reserves and gave suggestions for reducing the project's cost, such as making the proposed outdoor enclosure smaller. Captive breeding was considered, the laboratory almost ignored. And Sir Peter concluded, "The emphasis should be on field work."

Wang Menghu, single-minded in pursuit of China's interests, reiterated what he had said. By turns voluble, humble, witty, forthrightly aggressive, and bland, he played to his audience of colleagues as much as to us. "A research center is needed to coordinate all panda work. Without such a center, other units will not cooperate. . . . The first step is the construction of a research center." His message was clear: no research center, no panda project.

"WWF raises money from the public," Sir Peter replied. "It is not a bank. The public does not like to pay for facilities. At present, WWF cannot agree to pay for capital construction."

On that tense note the session ended.

WWF began the project with romanticism and naiveté. Foreign businessmen in China think in terms of customers, as missionaries once thought in terms of souls. By contrast, WWF conveyed that it was here solely for a selfless purpose. Was not a shared commitment to the panda enough? Other countries had accepted with gratitude whatever WWF had offered in terms of funds and equipment. The Chinese have long perceived that foreigners always serve their own interests, whether they are bestowing gifts or taking something. The Chinese do not confuse material interests with other goals. At any rate, WWF was unprepared for hard negotiating.

WWF lacked its own interpreter, someone to explain its position in detail, to catch nuances, to provide impressions. Instead it depended on Ma Huanqin of the Environmental Protection Office. A cheerful woman with a radiant smile, Ma Huanqin worked with sincere dedication to bring us together, but of course her obligation was to the Chinese. She even gave up her *xiu xi*, the two-hour noon rest that is a worthy but inefficient Chinese institution, to promote our work, and in late-night sessions translated what we had written. But an interpreter's task is an onerous one, and her translations sometimes caused confusion and agonizing misunderstandings. With the benefit of hindsight, it is also clear that WWF made and continued to make the error of sending its top persons to bargain on even trivial matters. Negotiations are handled by those of middle rank in China; Wang Menghu's immediate superior never appeared at our sessions during these days, and I did not meet any high leader in the ministry for years. Also, if a top person negotiates there is no face-saving way to withdraw, deflect blame, or defer decisions. WWF unconvincingly used its impersonal executive committee for that purpose.

Only two days had been budgeted for discussion and agreement on the complex issue of a research center, just before Sir Peter's planned departure. The Chinese had deftly manipulated WWF by creating time-pressure, an excellent negotiating tactic. They knew that WWF desperately wanted the project for its twentieth anniversary year in 1981; they knew that WWF had learned that Vice Premier Fang Yi had agreed in principle in November 1979 to a panda study by the Smithsonian Institution, raising for WWF the specter that another organization might be the first to study its symbol; and they knew that Sir Peter had to have an agreement now or delay the project, cancel fund-raising efforts, and disrupt other plans. Momentum was such that it would be easier for WWF to go ahead than to pause and consider its goals. Agree in haste and repent at leisure. Or, as Wang Menghu noted on another occasion, there is a Chinese proverb: "*Qi hu nan xia,* when riding a tiger it is difficult to get off." And he played on our nerves accordingly.

The next session was devoted wholly to finances. Wang Menghu spoke: "The Chinese side needs to know what percent of total construction cost World Wildlife Fund can provide because in China any project must submit a written plan to the State Capital Construction Commission. Plans must include an objective, the size of the facility, the cost, and a timetable of construction. . . . No plan can be submitted until the World Wildlife Fund contribution is known."

He wanted a figure now. Today.

Our sessions included a doughy and dour Party official whose main contribution, it seemed, was to provide subtle pressure with muttered asides, such as, "We don't need foreign help. We can do it ourselves." As he was well aware, Nancy Nash could understand Chinese.

"World Wildlife Fund has no money for the project at present," replied

Sir Peter. "Perhaps only ten percent and certainly no more than twenty-five percent of the funds raised for pandas could be used for construction."

Such vague figures did not satisfy the Chinese side. Wang Menghu suggested that WWF contribute fifty percent of the construction costs. Sir Peter countered that the cost of the research center should be cut to one million dollars. The Chinese knew precisely how much they wanted; WWF had failed to decide on a bottom line, an awkward negotiating tactic.

That evening Sir Peter telephoned Charles de Haes in Switzerland. For reasons unknown, the receiver was amplified so that his words could be clearly heard by all of us in the room.

"One million dollars doesn't scare me. Or even two million," said Charles de Haes.

We were delighted that an agreement on the panda project could now easily be reached. Charles de Haes's personal commitment on behalf of WWF would make the research center a reality.

The next morning, Sir Peter opened the session. "It is recognized that the research center must be built for the protection of the giant panda. . . . World Wildlife Fund International is prepared to make available to the People's Republic of China the sum of one million dollars. . . . The sum does not include equipment, field research expenses, overseas travel, and other such items."

I expected happy smiles or at least a visible relaxation of tension. But Wang Menghu, Hu Jinchu, Ma Huanqin, and the others just sat, pretending to be utterly nonchalant. Only Xiao Yu, a tiny sparrow of a woman, committed herself to a visible reaction: her pen remained poised above her notepad instead of scribbling down busily whatever anyone said.

"One million dollars is fine," answered Wang Menghu laconically. "When will the one million dollars arrive?" Business was business.

Finances having been more or less settled, the Chinese now wanted an action plan, a timetable for construction, and a five-year schedule of research on pandas in the wild and in captivity. By tomorrow morning. Sir Peter, Nancy, and I discussed and wrote far into the night and Nancy also typed various drafts of the plan.

After Sir Peter left Beijing, Nancy and I worked yet another four days on the details. The Chinese would take our English text, make a rough translation into Chinese, and then convert that version into English, an exercise that left many sentences unintelligible. Word by word, sentence by sentence, we would go over drafts, everyone arguing over subtleties of meaning. We thought we were working out a final version. But others in the bowels of the bureaucracy would then make further revisions, often incomprehensible and sometimes sly. It was for instance agreed that WWF would provide one million dollars for construction; someone changed this to "at least one million dollars." Such mischief kept me alert.

Other matters, too, were considered. An expert in the management of captive animals was needed immediately to work with local architects on the design of the research center. WWF had stipulated that it would have to approve the design before signing the action plan. Buildings would have to be traditionally Chinese, with the lovely curving roofs that prevent ghosts and demons—who can travel only in straight lines—from entering. William Conway, general director of the New York Zoological Society, was willing to assist, but he needed a visa. The Chinese can move swiftly when necessary, and in this instance we were helped by Wang Jinhe, a *Xinhua* reporter who had attended several of our sessions. A visa was quickly arranged. I met Bill in Chengdu and we spent several days in Wolong. The breeding facility, it was decided, would consist of a quarantine unit, a nursery, a service unit, and several individual panda enclosures. Each enclosure would have a shelter that provided the animal with privacy and an outdoor yard connected at the far end to all other yards by one long enclosure. This would enable a female in estrus to become acquainted with each available male or vice versa, the animals selecting a compatible partner by themselves.

Another item was my schedule. It was agreed and so noted in the action plan that my work in Wolong would begin in November of 1980. Zhang Shuzhong, a reserved and watchful old man with wisps of white hair, hinted that life in Wolong would be too difficult for foreign women. Chinese avoid confronting anyone with something unpleasant, but his message was clear: my wife, Kay, would not be welcome. I had already learned that little would be gained by discussing such a matter at length in a formal meeting, where even a simple problem becomes a morass of complications when subjected to a dozen viewpoints. Instead, I told Wang Menghu informally that I would not work in China without Kay. Having gauged my inflexibility on this matter, he conveyed that the problem would soon be resolved.

The action plan was completed. The meetings were over. William Conway assessed the broad impact of the agreements in a letter to Charles de Haes: "It is clear that WWF has struck a remarkable bargain. Its modest contribution to the giant panda program and related developments, even should this amount reach several millions (as I hope it will), is having a disproportionate effect upon conservation consciousness world-wide. . . . It may well be that nothing WWF has done heretofore is so truly international in its implication for wildlife conservation."

The protocol and action plan were signed by Li Chaobo, the head of the Environmental Protection Office, and John Loudon, the president of WWF, on 30 June 1980 in the Netherlands. The first words read:

> The Chinese Society of Environmental Sciences of the People's Republic of China and the World Wildlife Fund International have recognized that the Giant Panda is not only the precious property of the Chinese people, but also a precious natural heritage of concern to people all over the world. It has great scientific, economic, and cultural value.

We were soon to learn that the signing of these agreements signaled only the beginning, not the end, of negotiations—and of serious problems. Nevertheless, half a year later, in December 1980, I was back in Wolong to join Hu Jinchu and begin our detailed study of the panda.

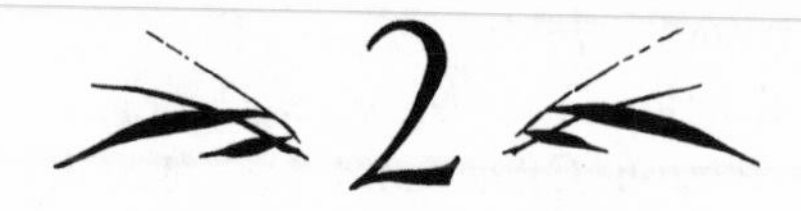

Winter Days

December 1980 to February 1981

Although it was still dark at seven in the morning, I heard Tang Gianrui, the cook, clanking around the kitchen shed. I pulled my shirt into the sleeping bag to warm it and then lay rigid until its chill had passed. Hu Jinchu was on the other side of the tent softly snoring, and at the far end slept Xiao Huang, a temporary interpreter from Beijing, wearing his ever-present fur hat. With dawn I slipped into my clothes, my Germanic work ethic making it almost a sin to remain in bed after daybreak. The occupants of the other two tents of our Wuyipeng camp were still deep in slumber, and outside the gray sky spoke of another cold December day. The thermometer, attached to the trunk of a rhododendron, registered 18° F.

I ducked into the communal shed. A monument to simplicity, it consisted of a ramshackle collection of rough-hewn boards, their many gaps partially covered by grass mats, and a roof of tarpaper. Canvas hanging over the door prevented some of the wind and snow from drifting through. It was cold inside, that is the best one could say. Tang Gianrui, looking like a burgundy-colored bear in bulky down jacket and pants, hovered over a wok and a pressure cooker preparing breakfast. The hind quarters of a pig hung suspended from a rafter, and in a corner were sacks of flour and rice, containers of cooking oil, and other provisions. Taking a *mantou,* the bun-like steamed bread that is a Chinese staple, I stepped into the adjoining room and placed the *mantou* by the hot coals of a wood fire that burned in the middle of the earthen floor. Sitting on one of the wooden stumps that served as seats, I watched the *mantou* toast and warmed my hands over the fire. The room was bleak, only marginally livable. In one corner was a barrel surrounded by a small glacier of frozen spilled water. Every day Tang Gianrui filled the barrel with buckets of water carried with a shoulder pole from a rivulet several hundred feet away. In another corner was a rickety table with a transistor radio, and a few shelves on which everyone kept his own cup stuffed with washcloth, soap, toothpaste, and toothbrush. Firewood was stored in the low rafters, black with soot like everything else. There was no window, but one wall had a large opening to permit logs to be pushed directly onto the fire from outside.

Zhou Shoude appeared. Taking one of the enamel washbasins that littered the floor near the barrel, he dipped warm water from a pot suspended over the fire and washed his face. He was camp leader, a trim young man, always civil but with a distant manner that isolated him from me. Zhang Xianti came and stirred the logs, making sparks fly. His hair was short, dense, and upright like the pelt of an electrified mole, and his smile was wide on a broad face. Always helping where needed, he was a methodical worker. Tian Zhixiang then joined us, a charming superfluous man whose good humor and friendliness more than compensated for the fact that he was one of those people of comic incompetence who cannot put a foot quite right. Having failed in his logistic job of keeping Wuyipeng provided with ample food, the authorities appointed him to do research. More of the team came, Pan Wenshi and Xiao Wang and Da Wang—Little Wang and Big Wang, as we called them—until all eleven of us were there. "*Chi fan,*" called Tang Gianrui, "Food." We lined up to ladle rice gruel into our enamel bowls, then added roasted peanuts and pickled vegetables. Huddled around the fire, we ate hurriedly with little social chatter.

"Where shall we go today?" Hu Jinchu asked me.

Every day so far we had gone to a different part of the Choushuigou to help me become familiar with trails and terrain. The drainage is small, only about ten square miles, and the main valley divides into three forks. Many ridges project into the drainage, creating a rugged topography of narrow crests and steep slopes. Only on the side opposite Wuyipeng is there a moderately gentle plateau named Fangzipeng.

"You decide," I replied.

I admired Hu Jinchu's dedication to field work. By the age of fifty, most of the biologists I met in China had mentally retired. They did little beyond protecting whatever status they had from young aspirants, and keeping a low profile until a pension was in hand. To this, Hu Jinchu was a happy exception.

Hu Jinchu had selected this camp site in 1978 and knew the area well. Located at an elevation of eighty-three hundred feet near the mouth of the valley, Wuyipeng perched halfway up the slope on the crest of a spur, in the transition zone between two bamboo species. We could go down into umbrella bamboo, or walking-stick bamboo as the Chinese call it, or up into arrow bamboo. Hu Jinchu decided that Peng Jiagan would accompany us to the northern fork of the valley. Another team would sample bamboo—count and weigh dead stems, young stems, and old stems in plots of one square meter. This would give us an idea of stem density, stem production this year, and the amount of food—the biomass—available to pandas. The Chinese wound long strips of coarse wool, puttees, from ankle to calf for protection against wet and cold but wore only flimsy tennis shoes. I made a mental note to obtain boots for the team.

The trail traversed the slope and Lao Peng—*Lao* is an honorific meaning "old"—was in the lead, wearing his usual goatskin vest, hair inside. He moved over the uneven ground with the efficiency of someone used to mountains, and indeed he had once been a hunter here and then a warden after a small portion of what is now Wolong received protection in 1965. Although snow covered this shady slope, the forest seemed strangely verdant. Beneath scattered spruce and hemlock were rhododendrons, their leaves curled tightly against the cold, and large stands of bamboo in green leaf as if summer had been momentarily suspended. We soon passed *Bai Ai,* White Rock. From the top of this small cliff one could see beyond the Choushuigou northward across ridge after ridge, those nearby sharp with a crystal clarity, those distant faint gray brush strokes upon the sky.

Just beyond Bai Ai a trail plunged down. Grasping bamboo stems and saplings to slow our momentum, we careened to the bottom where a brook tumbled among boulders glazed with ice. We crossed it on a bridge of two slender logs. Our trail now ascended to the crest of a ridge and we halted there briefly, breathing heavily. Down the other side we came to a seepage smelling of sulfur where the ground had been churned by the hooves of takin, ponderous relatives of the Arctic muskox. Lao Peng told us that in summer the takin come here from valleys to the west to eat the salt-soaked earth. A waterfall blocked our path and we had to inch along a ledge to reach a broad, slanting ice chute by the falls where a slender log spanned the ice at an angle. By clutching the log and balancing on ice steps hacked by Lao Peng's curved knife, we climbed up and across the obstacle, chill water foaming in a pool below.

Willows lined the stream's edge, and a white-breasted dipper fled upstream, skimming the water. Lao Peng halted abruptly by fresh panda tracks only hours old. These tracks seized my imagination and transformed the mountains as I looked around, almost expecting to see the

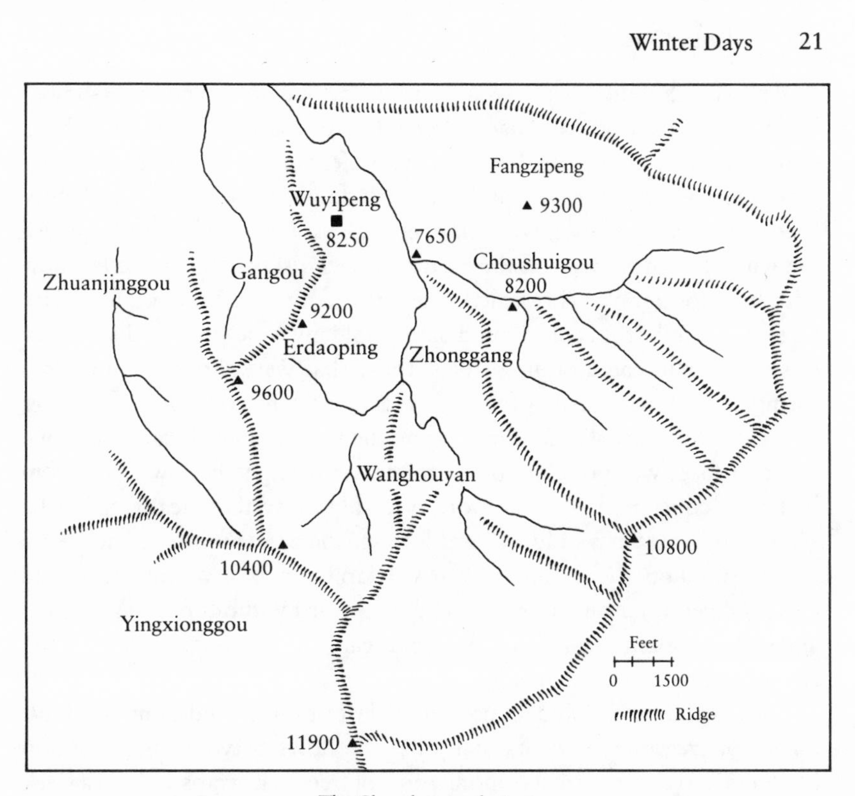

The Choushuigou drainage

haunting image of a panda moving silently in the bamboo. But there was no stir, no glint of black and white. Yet the forest seemed aware, watching, as we trailed the animal. The tracks meandered on, toes pointed inward, strangely blurred as if the animal wore booties of burlap, the result of its hair-covered soles. At times the panda waded across the stream and at those points its footprints were briefly glazed with ice. Two droppings nestled in the snow like eggs in a nest, last night's, for they were already frozen. I slipped them into a plastic bag to take back for analysis. We collected at least thirty droppings every month, dried them, separated leaf from stem—an easy task because pandas barely chew their food—and weighed each portion to determine if the animals seasonally preferred certain parts of the bamboo.

The tracks veered up a steep slope. Well guarded by the dense bamboo, the panda did not reveal itself. Besides, Lao Peng talked in a loud voice and our steps were excruciatingly noisy. But by examining the trail, we could at least learn some of the panda's secrets. Squat and barrel-shaped, the panda was ideally designed for passing under downed trees and snow-bowed bamboo. We were not, and had to bend the stems aside to pass through. At about five feet, arrow bamboo was the perfect height to hinder visibility, whip faces with wet leaves, and sift snow down sleeves and neck. Handholds gave way as brittle twigs snapped and our feet slipped from logs. All the while I tried to pace off the panda's travel distance. Compacted snow with bamboo remnants revealed where the animal had sat and fed. With fingers so numb from cold I could barely hold my pencil, I recorded the age of stems it had eaten. Young stems, those that were shoots the previous spring, lacked the branches found on older stems. By counting the number of each stem type eaten and measuring the length of each stump and discarded top, it was possible to determine approximately how much stem had been consumed. Tedious work, yet only by systematically gathering such mundane facts would we be able to see the panda's life with new eyes.

We halted on the crest to eat our spartan lunch of *mantou.* The panda had followed the ridge, moving steadily upward without pausing to feed, and soon we lost its trail in a snow-free area. It was mid-afternoon, time to retrace our route to Wuyipeng.

Back at camp I checked mousetraps. The terrain around camp was ideal for small creatures, with its many dark hollows between moss-covered boulders and secret retreats among roots of trees. The traps yielded a sleek tan mouse resembling a deer mouse, and a shrew with a peculiar proboscis, wrinkled and bristly like an elephant's trunk. Hu Jinchu identified the former as belonging to the genus *Apodemus* and the latter as *Soriculus.* I gave him the frozen bodies for museum specimens. On projects in the past, I had collected everything from plants to skulls; I was a hoarder of small treasures. But as I grow older, such things distract me less, except perhaps for a note or sketch in my journal. Also, the Chinese considered collecting by foreigners a form of scientific imperialism. Western museum collectors had once plundered archaeological treasures, collected plants and animals at will, and bestowed scientific names on most of the country's species. I was not permitted to collect anything except what was directly related to pandas.

The hut was already crowded as I sought a seat, trying to blend into the companionship of the group. I regarded China as a challenge, and my arrival at Wuyipeng had been greeted in the same spirit though with different perspective. Most of my coworkers had never seen a foreigner before 1980, and none had ever associated closely with one. Pan Wenshi from Beijing University and Wang Xuequan, to give Da Wang's full name, from the

Northeast Forest Institute in Harbin were from large institutions, but with the exception of Hu Jinchu, the staff consisted mainly of Sichuan workers with only a basic education. I was exotic and probably bizarre; I certainly needed to be broken of some barbaric habits. I said "Thank you" when someone helped or showed kindness. Such casual thanking was a cheap payoff; one does not express gratitude with words but with later deeds. I greeted others in the morning and said "Good night," behavior quite "ridiculous," as one expressed it. Why was I so infernally formal? Not for months did I learn that these were mistakes, but in other errors I was more immediately instructed. Upon return to camp in the evening, for instance, I was expected to wash my feet in a basin of warm water. And I did.

Then came dinner—a bowl of rice topped with stir-fried cabbage and bits of pork fat. Food was simple but good, better than the meals of the average Chinese, for we had a little pork every day. Key food items such as rice and cooking oil were still rationed, available only with coupons. Chopsticks clattered on metal bowls like rain on a tin roof as we ate. Da Wang, a tall exuberant chap, slapped a ladle of rice in my bowl after I had twice declined more. I did not want it. I felt defiant, but going overseas imposes a duty to fit in, and I passively ate. I hoped that friendliness would with time become less aggressive, smiles less anxious.

By seven in the evening the hut was again dark except for the flickering fire. There was only one functional lantern, a small kerosene one, and the cook had it. I went to our tent, where Hu Jinchu had started a fire. Smoke leaked from every joint of the stove as the wood barely smoldered; the stove pipe needed cleaning. But by lying in bed I was below the sheet of smoke and could breathe while transcribing the day's notes by flashlight. A research tent was needed, a clean, dry place to keep equipment, specimens, and communal notes. We had prepared forms on which to record data, and these Hu Jinchu stored beneath his pillow, together with books, correspondence and other papers. He slept on them, safe storage but not convenient.

Another day. Frost covered the birches around camp, each twig shining like transmuted starlight against the morning sky as Da Wang and I left to seek pandas on a ridge above camp. I liked Da Wang. Vigorous and intelligent, he added sparkle to our restrained group. Someone had confided that his job was to keep me under surveillance. Whether this was malicious gossip or true, it did not disturb me any more than the knowledge that my mail was being read, coming and going.

The trail assaulted the slope with a frontal attack, straight up. Beginning with a rocky pitch so steep we had to pull ourselves up by grasping trees, it then followed the top of a ridge, surmounting two giant steps before leveling off among huge firs, some three or more feet in diameter. After a while the trail led to a small plain named Erdaoping, where arrow bamboo covered the slope in one vast thicket so dense that little but moss could grow in the shade beneath. Only where a tree had crashed, breaking the canopy of bamboo, could new shrubs and saplings reach light and survive before the bamboo engulfed them. The trail had been cleared with machetes, leaving pointed bamboo stumps, an army of tiny spears ready to impale any traveler, and adding considerable incentive not to fall, no matter how slippery the trail was with mud or snow or ice, and often all three. With time I knew every handhold and foothold along that trail and every unstable log and deceptive patch of earth; I knew even the feel of the bark of certain trees.

After about an hour's climb we reached the ridge. Here at ten thousand feet it was cold, but we were liberated from the valley, a horizon suddenly offering visual release. Mountains sparkled in air so clear that it seemed I could hurl a rock into the Pitiao River far below. This ridge overlooked the small Zhuanjinggou and the Yingxionggou where the breeding station was located. On the other side of the Pitiao valley was a grass slope with only an upper fringe of trees as remembrance of forests past. Beyond, at over twenty thousand feet, the ice pyramid of Siguniang held itself aloof from the hills. Siguniang means Four-Maiden Mountain, and legend has it that one day four maidens joyfully climbed its slopes. They decided to remain forever, and you can still hear their laughter and see their lovely shapes. For years their mother called them, begging them to return, but they preferred the company of clouds, the mountain peaks of white jade, and below them the yaks on meadows of wildflowers in bloom. We followed the ridge as again and again it lunged upward. The forest was silent yet full of sound, with iced boughs tinkling like temple bells in gusts of wind.

To our left in the Choushuigou was fir forest; to our right in the Zhuanjinggou, a slope with but a few tattered birch, the area denuded by lumbermen in 1972. It was a reminder that Wolong, no matter how seemingly remote, had not escaped the heavy hand of humankind. A field biologist is cursed with the ability to see in ways that others often do not. In the Choushuigou there were many places whose beauty I could not appreciate because the original forest had been destroyed. Near Wuyipeng was scrub rather than tall spruce piercing the fog; across from Wuyipeng was a larch plantation that replaced fields, some say of opium poppies; and the birch forest at Fangzipeng had lost its fir, and only old stumps attested to their once-grand size.

Over four thousand people now live in Wolong, most of them downstream near the reserve entrance where the Pitiao joins another river to become the Gengda. Many of the residents belong to the Qiang tribe. In the past, until about the end of the Tang Dynasty in 906 A.D., the Qiang were a powerful and widespread agricultural people who lived to the northeast in parts of Gansu, Shaanxi, and Qinghai provinces. Defeated in battle, many are said to have migrated into the mountains to the southeast. There they encountered another tribe, the Ko-jen. A divine dream revelation told them to fight the Ko-jen with clubs and white stones. To commemorate their victory, a white stone, usually a piece of quartz, became the Earth God, and it was worshiped in sacred groves of trees. Until recent decades, villagers still placed a white stone on top of their homes. Many Han Chinese settled in Wolong during the past century, the mixing of the two cultures and intermarriage diluting the traditions of the Qiang.

The Gengda River passes through a long gorge so difficult that travelers have preferred to detour across the mountains to the east. One route led up a valley near Hetauping, site of the research center, past a blunt peak named Niutoushan, Cowhead Mountain. Another traced the northern edge of the Choushuigou over a track that is still visible though much overgrown with bamboo. The few Westerners to reach Wolong used those routes. First, in 1904, was A. Hosie, who counted forty-six homes and five inns between Hetauping and the head of the Pitiao Valley. Ernest Wilson, greatest of the Western botanical explorers in China, came in 1908. While ecstatic about the flora in Wolong, finding "no region in western China richer in woody plants," he was less happy with the "execrable" trails. He noted that villagers carried logs out of the valley; even then timber was not just for local use in house construction as beams, boards, and roof shingles. After a road was built into the valley in the early 1960s, timber cutting for export to the lowlands increased dramatically.

Da Wang and I failed to find recent panda spoor that day, a situation all too typical. Where were the animals? Some ranged out of our valley, and others were perhaps so sedentary that we did not cross their tracks. Whatever the reason, pandas were scarce, much scarcer than I had expected. After all, they had an unlimited food supply, they were literally engulfed by bamboo. Having searched all day, there were evenings when I returned to camp with my clothes sodden and body chilled, my knees aching, my notebook virtually empty. I was obviously less resistant to cold and thickets than a panda. One point was soon insistently clear to me: the essence of panda tracking was discomfort.

Television shows and magazine articles have given the impression that field biologists lead an exciting life as they wallow with whales and associate with amiable apes. All but forgotten is the fact that many creatures are solitary and rarely seen, and that self-imposed isolation means lack of conveniences, bewildering cultures, and penance in dust, heat, snow, or rain. Most people would find little romance in field work. One night, withdrawn in my sleeping bag, I mused to myself that a field biologist's greatest danger lies not in encountering fierce animals or treacherous terrain but in finding comfort and being seduced by it. But there was no danger of that in this icebound, soot-blackened tent, I thought, as I contentedly drifted into sleep.

Even if pandas remained elusive, we still had small adventures that interrupted the day's routine. Leopards occasionally patrolled the trails, and we found their droppings and tidy, round paw prints. Hu Jinchu told me that leopards sometimes kill young pandas, so we carefully examined the content of each dropping to discover what had been eaten. Most contained the coarse, gray-brown hair of the small tufted deer, and a few the hair of musk deer, golden monkey, and other prey. One day Pan Wenshi was ambling alone down a trail when he saw a leopard resting in a patch of sun. Both bolted. "I am much afraid!" he excitedly told me in English. Like a dragonfly darting over the surface of a pond, Pan Wenshi had previously done varied research, all in the laboratory. His enthusiasm seemed mainly reserved for such matters as determining the amino acid content of bamboo with high-pressure liquid chromatography, not counting bamboo stems on winter slopes. But I misjudged his interest in pandas. Later he would initiate and conduct China's best panda research program, not in Wolong but in the Qinling Mountains of Shaanxi. He would leave us in a few days, perhaps to return in spring, and I would miss his congenial presence.

Another day, after Da Wang and I had lost a panda trail, we meandered somewhat aimlessly toward camp. Coming upon a small cliff, I checked its base to see if perhaps a serow—a stocky, coarse-haired goat-antelope with stiletto horns—had rested there; animals usually centered their activity around such sites. Indeed there was a large latrine of elongated pellets to mark many visits by a serow. But more intriguing was a cleft slanting into the rock. I squeezed in. After about twenty feet, the ground leveled into a small chamber, and in it was a nest of conifer, rhododendron, and spiny

Berberis twigs. Lining the nest cup were chips of rotten wood bitten from two logs a panda had carried in. We had found a maternity den where, dry and secure, a panda female had years before raised a young during the first critical weeks of its life. I noted with interest that pandas could build nests. In our tracking we had found many panda rest sites, easily recognized by the packed ground and the many droppings around the perimeter. The animals had merely reclined on damp earth, snow, or frozen ground; body heat had sometimes melted snow into ice. Yet pandas obviously sought some comfort, for they often selected a rest site with something, such as a log, to lean against. The base of a big fir where the ground was soft and dry with humus was also favored. So why not build a nest? A gorilla would do so. And, to my surprise, the Asiatic black bears do too. These bears resemble American black bears except that they are chunkier, shaggier, and have a large white chevron on the chest. They were scarce around Wuyipeng, where I found just three of their nests, each of bamboo stems bent and folded to provide a soft bed. Perhaps the panda's coat, dense, woolly, and with a springy quality, is such that added comfort is seldom needed.

In a forest where most large mammals were devoted to remaining hidden, the golden or snub-nosed monkeys were an exuberant exception. When the French biologist Alphonse Milne-Edwards first described the species in 1870, he named it with Gallic humor *Rhinopithecus roxellanae* after Roxellana, the beautiful snub-nosed mistress of a Turkish sultan. The fate of the golden monkey is tied to that of the giant panda, for their distributions are almost identical. About two hundred to three hundred monkeys ranged through these mountains, visiting the Choushuigou only intermittently. Sometimes I neither heard nor saw them for weeks, then suddenly a golden horde would descend into the valley. Dead birch branches snapped and crashed as the monkeys leaped from tree to tree, and nasal wails like those of a petulant child kept scattered group members in contact.

The monkeys were shy, for they had been heavily hunted and their skins were still for sale in the fur shops of Chengdu. During the Qing Dynasty (1644–1911), only high officials were allowed to wear clothing made of golden-monkey skins. In the 1930s, a golden monkey pelt would bring the large sum of two hundred dollars in the local fur market. Every Qiang used to have a personal guardian spirit that manifested itself in objects or other creatures, especially in the golden monkey. Priests often carried the skull, liver, fingernails, and other monkey parts wrapped in white sheets of paper in remembrance of the tribe's sacred books. Once, long ago, when wars forced the Qiang to emigrate, their sacred books became wet as the people crossed a river in leaking cowskin boats. As the books were being dried on shore, a goat ate them. A wise golden monkey appeared and advised the people to eat the goat and make a drum with its hide. The beating of the drum would help everyone remember the written word.

We often tried to observe the monkeys but were quickly spotted, often by a peripheral male who would look at us intently, then give a chucking call like that of an agitated blackbird. If we stood still, the monkey might tarry, his bright blue face inquisitive, a wart-like growth of unknown function in each corner of his upper lip. All too quickly, however, the male would hurl himself from the branch and wildly leap away trailing his mantle of golden hair. And the others, startled by his flight, would follow.

It had been decided that at least one Chinese must accompany me at all times, such solicitude being based on fear that something might happen to a foreign guest. Besides, Chinese are determinedly social and generally do little alone that can be done together. To me a crowd is anathema; I like to be surrounded by sufficient space. In this project more than any other, there was always the conflict between what I wanted to do, what I should do, and what I was permitted to do. I resolved the problem of a field companion by asking to go on short lone walks at a time when no one would want to accompany me, as at day's end, and soon no one questioned these solitary ramblings. On one such excursion I found a group of sixty to seventy golden monkeys, most sitting peacefully on spruce boughs in the sun that had just descended the slope. I also settled myself in the warm rays, my back against a tree trunk. To one side, four males were in a birch stuffing themselves with lichens that festooned the branches; they plucked lichens either by hand or first broke off a branch and then, holding it, leisurely nibbled them off the stick. Seven juveniles clambered playfully together among the adults. Near me were several females with small young. A large male swaggered toward them, his tail looped in a high arc, showing his dominance. Approaching a female with infant held to her chest and a juvenile at her side, he tenderly groomed the latter. Then going to another female, he tried to groom her infant, but she rebuffed him, holding it tightly. Forcefully he pulled the infant from her, groomed it gently, and, having asserted himself, he just as gently released it back into her arms.

Observing these golden creatures and taking notes on their behavior was a pleasure all too rare. Once, when I tarried with a monkey group,

Zhou Shude reminded me: "You are here to study pandas, not monkeys." Our work was limited to what was specified in the official action plan.

On yet another day, several of us crossed the valley to Fangzipeng. Along the plateau's edge, where it drops into another valley, I noticed dead branches arranged to make a crude fence. Having seen similar fences in other countries, I knew it had been built by poachers. The fence soon led us to a gap with a wire snare in it; one end of the wire was tied to a leaning sapling, the other was a suspended loop. An animal would step into the opening unaware of danger until it felt the noose around its neck. Filled with panic, it would lunge ahead, rather than back away, and the noose would then tighten, cutting into its neck during the violent struggles, until death would come slowly by strangulation. Hu Jinchu told me that poachers here were mainly after musk deer, for the male musk deer has a gland on his abdomen that contains musk, a valuable commodity whose supposed virtues are highly prized in oriental medicines; it is the living equivalent of gold. Used in over a hundred medicines, musk is said to cure asthma, typhoid, impotence, pregnancy, apoplexy, pneumonia, and other problems. I had laryngitis once and was given a box of vials with a total of about three hundred black pills manufactured in Chengdu. The pills had these listed ingredients: "Rhinoceros horn 10%, pearls 7%, bear gall 7%, cow bezoar 5%, musk 3%, toad cake 10%, rhizoma coptidis 28%, fined borax 30%." I was prescribed thirty pills a day and soon was cured, though whether by time or the medicine remained unclear. That day we collected eight snares. A panda had blundered through the fence within a few feet of one of these. Hu Jinchu delivered the snares to headquarters, and I hoped he would object strongly to their presence in a reserve. But I had already noted that the years had trained him in an important survival skill, to suffer in submissive silence, to ignore rather than protest. Consequently in subsequent meetings I emphasized the need for patrols to protect wildlife—but to no avail. The police force in Wolong was too busy with other duties, I was informed, though it was my impression that their daily task consisted chiefly of reading newspapers and drinking tea.

I had been in Wuyipeng only a month, yet I went into the field each day not in peaceful pursuit of knowledge but under bleak mental pressure to amass quickly as much information as possible. The work was going well in that we were sampling bamboo and determining the panda's food habits. Details of behavior would have to await radio telemetry. Pandas would have to be caught in live traps for radio collaring by April so as not to disrupt any mating activity. It was already mid-January, and Howard Quigley of the University of Tennessee was due to arrive at the end of the month with the radio transmitters, yet only one trap had been built so far. Would Howard, and Kay with him, be permitted to come? Would the project in fact continue?

We all knew that the WWF-China meeting in November had failed to agree on equipment for the new research center. Zhang Shuzhong had then written a letter stating that "if this problem couldn't be settled ahead of time, the field research which though in process [*sic*] should be stopped temporily [*sic*] from Feb. 16th, 1981. . . ." Another meeting to decide the fate of equipment—and our work—would be held in mid-February.

My Chinese colleagues seemed detached about this matter. For them the collapse of the project would be a deliverance, not a loss. Everything had been disrupted by my arrival: the chain of command, the methods of work, and for several their usual jobs. Da Wang would quickly return home, for "I am heartsick for my son," as he put it. Hu Jinchu could continue his survey of Sichuan's fauna. Not one person in camp except myself was there because he wanted to work on pandas. The others had been ordered to do the task by the Party and the bureaucracy; for most the fact that they were in Wolong indicated that they lacked the political contacts to be elsewhere. Potential interest shrivels under such compulsion. How would the research center ever attract competent staff when almost everyone applied for a transfer as soon as he or she was sent to Wolong, where the weather and living conditions were foul? No matter how irritated and dispirited I became because of the gap between promise and performance in the project, I tried to remember this fact: love for the study was my primary motive—and only mine. Realizing this, I was not only passionately involved, but also became more compassionate, or at least I tried to be.

A message was delivered that Howard Quigley and Kay would arrive in early February. I was relieved that the Chinese side had at least tentative plans for the project's future, and eagerly went to meet the two in Chengdu, happy to see Kay and pleased to greet Howard. Howard had written me almost a year earlier to express interest in joining the panda project. He was just finishing a thesis on radio tracking black bear and, he noted, he was considered to be competent, industrious, and congenial—an evaluation I found wholly accurate. Unlike most applicants, who seem to take for granted that love of animals is the only requisite for a field study, Howard had experience. Never having sedated a panda or radio tracked any animals except lion, puma, and jaguar, I felt it best to have someone knowledgeable about bears teach us the relevant techniques. Howard would join us for two months and then take my place on a jaguar project I had established in Brazil.

We stayed at Wolong headquarters for a night on our way to Wuyipeng. The cement walls of the unheated rooms radiated cold. We could request either a small iron bowl with burning charcoal or an electric hot plate, both offering enough heat to at least warm our hands. Because no one had drained the water from them, ice had burst the toilet tanks, new only last summer. The room's single light bulb glowed with five-watt power, for the hydroelectric station was almost inoperative due to the river's low winter water level. How could delicate equipment in the research center function properly under such conditions? The only comfortable place was in our sleeping bags; I longed for the smoky, open fire in the hut at Wuyipeng.

There had been excitement here the day before. Four Asiatic wild dogs—fox-red, pack-hunting canids—had pursued a panda that had eluded the predators not by climbing a tree, as I would have expected, but by fording the Pitiao River. The hunt was observed by local people, who herded the exhausted panda into a sheep pen. From there it was transferred to headquarters. Bi Fengzhou asked if we wanted to see the animal. Slight and seemingly frail, Bi Fengzhou had the energy and tenacity to make things happen; he was then our chief contact at Chengdu's Forest Bureau, and we all appreciated his kindness and depended on his help. We peered through a barred door at the panda. It was a female, old and lean, the skin taut over her broad skull; her two lower canines were mere stumps. Six of us crowded within three feet of her, yet she placidly took sugarcane from our hands and ate it, indicating by neither sound nor action that she had been wild and free only yesterday. The windows on her emotions seemed closed: her eyes were tunnels without light, her ears perched like lifeless decorations, her face was mute and diffident. Her years were almost ended; old and weak, she would fade away. I suggested that she soon be released back into the forest. No, I was told, she would be sent to the station at Yingxionggou, a fate which inspired my pity and concern.

Over a month earlier I had revisited Yingxionggou to see whether William Conway's suggestions for improving panda management had been implemented. He had recommended cozy wooden nest boxes; there were now bare boxes lined with sheet metal. He had recommended some protection from wind; there were now solid shutters that plunged the animals into perpetual night. Dark, cold, their cages encrusted with frozen urine and feces, the pandas suffered silently, curled up, eating little. One youngster had become ill and had been taken to Chengdu. What of the other youngster? It had died last summer, I was told, of starvation. A veterinarian specializing in chicken diseases had permission to study these pandas. Bing Ji or Sick Chicken as I shall call him, was a powerful Party official at a local college who could do as he pleased. And it had apparently pleased him to do feeding experiments on the baby panda, with tragic results. The Yingxionggou staff could not comprehend my furious determination to change things. These men were not so much callous as oblivious to the needs of the pandas. It did not occur to them that the animals were desperately unhappy. I found little empathy for animals in China. Perhaps noteworthy is the fact that the Chinese written character for "animal" means "moving thing."

Village women wearing dark cloaks of homespun and white turbans carried our baggage on the one-hour climb to Wuyipeng. Kay and I now had our own tent, isolated over one hundred feet downhill from the hut, in a small clearing surrounded by umbrella bamboo and birch. The inside tent walls were plushly lined with red blankets for insulation. At one end was a double bed and beside it a box used as a washstand; along the opposite wall stood a desk and chair, and in the middle of the floor we had a tiny stove. All was fitted into a ten-by-thirteen-foot space; it was so crowded that if one of us moved around the other had to sit. We pounded nails into the tent frame so we could hang equipment and strung rope on which to drape clothes. Yong Tianzhong had piled firewood outside the door. A quiet man with shoulders that a weight lifter might envy, Yong Tianzhong was a relentlessly hard and efficient worker as day after day he hand-sawed, carried, and chopped wood to keep the cooking and camp stoves going. Our tent was soon a cozy home. And finally I had someone to listen to my worries, to share insignificant thoughts, to reach out and touch. I could relax. I could share feelings and emotions with someone rather than just endure the forced companionship of camp where I usually began a conversation with the thought, "What answer will my comment allow them to give me?"

Toward dusk, Kay and I liked to stroll toward Bai Ai, moving quietly in the gloom, hoping to surprise some night creature or even a panda at a bend in the trail or in the darkness of a ravine. Usually there was nothing abroad, or at least we saw no movement except clouds spilling from the ridges into the valley and wisps of birch bark, thin as a snake's skin, quivering in the breeze. We examined the leavings of flying squirrels and a broken hydrangea twig whose bark had been gnawed by a golden monkey. Occasionally a Temminck's tragopan—a wine-red pheasant with a blue face—rustled away, head held low, to seek refuge in a tangle of brush. Or a musk deer might stand silent as fog, screened by a shrub, its miniature tusks gleaming, as it watched and waited to see if we would detect its presence. When we returned to the tent we still carried the scent of moss and spruce with us, and crystal beads of frost sparkled in our hair.

With Howard's arrival, our work settled into a new routine. Lao Peng was in charge of a team building live traps of logs. Each trap was about nine feet long, four feet wide and four feet high. The door hung suspended from the end of a pole, and at the other end a rope led to a simple trigger inside the trap. A panda pulling at the bait would dislodge a stick that held the rope in place, making the slide door drop. Meanwhile Howard and I concentrated on setting special snares like those used for catching bears unharmed in North America. One end of a cable is attached to a tree; then the other end, a loop, is placed flat on the ground around a small metal bar that projects into a little hollow. When it depresses that bar, an animal releases a spring that flips up and tightens the cable around its foot. Carefully we concealed cable and spring with dead leaves. To guide the animal into the trap, we built a V-shaped cubby of branches and logs that gave access to the bait from only one direction. And finally we tied bait to the tree trunk. Lao Peng and Hu Jinchu agreed that pandas like to eat meat above all else, so we used goat heads and pig bones. In addition, Lao Peng burned bones near the log traps, hoping the odor of roasting meat would entice an animal.

We soon had ten snares and four log traps in operation. Now we had to wait for a panda to pass and be tempted. With impeccable logic, the traps had been placed where pandas ought to pass, on ridges and trails and in valley bottoms. However, human logic is not necessarily panda logic. The traps remained disappointingly empty. But sometimes I found a trap door closed; after approaching cautiously, I would peer through cracks between the logs, wondering if I would see the quiet bulk of a panda, though usually I knew the answer. Silence differs in quality. A quiet snow night is unlike the expectant silence of a tense animal. Most often a weasel or yellow-throated marten had jerked at the bait, or a nutcracker—a jay-sized bird—had pecked at it until the door slammed down. Snares had to be checked twice each day, at dawn and dusk, and log traps once. It was hard, wearying work.

At noon, after checking the trap lines, we often gathered at the hut for lunch, a bowl of noodles spiced with soy, garlic, and red chili peppers. Our meals were not gracious events. Sitting in a semicircle around the fire, we sucked in the long noodles with swishing slurps. Noodles represent life—cut them and you shorten yours. The cold and the chili peppers made our noses run, which some emptied on the floor. Slurping, snorting, and sniffling, we had a musical meal.

After lunch, Kay and I were sometimes given a Chinese lesson. Chinese characters are essentially pictures, either direct representations or abstractions of objects, ideas, or actions. In spoken Chinese, a character or syllable has not only a certain sound but also a certain tone that determines its meaning. Being mostly tone-deaf and with perhaps an unreceptive mind, I had endless trouble with the four tones, each with its own inflection. With every word having at least four unrelated meanings, my verbal attempts were either received with blank looks or great hilarity.

"Táng, tāng, tàng, tǎng. Sugar, soup, hot, if . . . ," everyone would chant. "Máo, māo, mào, mǎo. Hair, cat, plentiful, rust . . ."

If I asked the name of an object, the answer created an immediate loud argument about proper pronunciation. The correct character would be traced with index finger on the palm of the hand with furious emphasis. The uproar always ended with the same conclusion: some had spoken in the Sichuan dialect, some in *Putonghua,* the official Mandarin. "Cup! Hand! Snow! Letter!" Da Wang would test me in English, saying each word explosively and then listening while I struggled with the correct tone until an "okay" signaled semi-success.

He Jin, a student at Beijing's Foreign Language Institute, had arrived with Kay and Howard for a two-week visit. Taking advantage of his excellent English, Howard gave a lesson on the use of the radio telemetry equipment—collar with transmitter, directional antenna, receiver in a waterproof case, earphones. He explained that the radio signal depends on line of sight: the signal cannot travel through mountains, and it is best to search for collared pandas by traveling high up on ridge crests. Everyone listened to the beep-beep-beep of the signal to learn how to distinguish an active from an inactive animal. A motion sensor in the transmitter sends out pulses of two speeds, one with seventy-five pulses per minute to indicate a resting animal, and one with one hundred pulses per minute to indicate a feeding, traveling, or otherwise active animal. By slowly moving the handheld antenna from side to side, Howard demonstrated how to determine the point where the signal is loudest, the direction of the panda. The actual location of the panda would become clear by simple triangulation, that is, obtaining directional readings from two to three points. He emphasized that signals may bounce from slope to slope, that they may fool the observer unless great care is taken. There was much else, lessons that could be absorbed only by actual experience. Whether we would be able to gain that experience depended on the crucial WWF-China meeting now in progress or soon to begin.

By the beginning of March 1981, we still had not received news about the February meeting in Beijing. However, our work had not been stopped, an auspicious signal.

So far the panda had been a creature of shadow rather than substance. For months it had permeated my mind; for weeks I had sought it in the forest. At times its presence seemed to vibrate in the air, but it had not once offered me a glimpse. Now our luck changed. Hu Jinchu fleetingly saw a panda, the first sighting by anyone in camp in over two months of field work. Surely I would meet one soon.

Panda Politics

During the first months at Wuyipeng, I had been tense and distracted, fearful that the panda project would be abruptly terminated by the Chinese. The signing of the joint agreement in June 1980 seemed festive. But by then mutual misunderstandings and repeated lapses in diplomatic etiquette and lack of sensitivity by WWF had created doubts about the project among the three Chinese agencies involved. A debate raged among them, some individuals wanting to cancel the project and others to continue it. As a compromise, it was finally decided to invite WWF to a meeting to discuss concerns and issues. This meeting was held in November 1980, just before I began field work.

I had planned to depart for China from my Connecticut home on 2 November because the action plan specified a November starting date for field work in Wolong. On 28 October, I received a telephone call from WWF saying that, because of problems, the Chinese side wanted to delay the project. Then, on the first of November, I was told to proceed to Beijing for meetings. It was an ominous beginning to the project.

The first meeting involved only WWF and the Chinese Environmental Protection Office (EPO). The WWF delegation consisted of Charles de Haes, Mark Halle, Nancy Nash, and myself; Mark's task was to handle WWF's "China desk." The China side had, among others, Qu Geping, a high-ranking member of EPO, and two interpreters, neither of them Ma Huanqin, who had been so involved with the project until June. The meeting was restrained. Among themselves, Chinese may exhibit strong passions, but during meetings with foreigners both sides are expected to be controlled and contained. However, this meeting was overwhelmingly cool. Qu Geping detailed various problems. Some were minor, the result of misunderstandings. For example, when a request is made by a foreigner, the Chinese, to whom a flat "no" is impolite, may smile and nod and say "yes" when in fact they mean "yes, we shall consider this," or "yes, we agree in spirit." When neither side is aware of such cultural subtleties, mixups are inevitable.

One major problem was publicity, or, as the Chinese say, propaganda. Qu Geping noted that WWF had been insensitive about the matter and had "hurt the feelings of the Chinese." In 1979, WWF had announced the panda project without involving *Xinhua,* China's news agency. Newspaper stories had given the impression that WWF alone would "save the panda." There had simply been too much wild beating of gongs and drums by WWF on its own behalf. More recently, WWF's insistent demands to make a panda film before the project had even begun also made the Chinese wary. Aside from the fact that they thought the request was premature, they also had become increasingly concerned about WWF's motives. Was the project mainly a commercial and propaganda venture to benefit WWF? Involved also, I think, was the self-effacing Confucian habit of avoiding acclaim, at least until a task had been successfully completed. I could understand WWF's desire to publicize the project. At least three other organizations had vied for a panda study, one of them promoted by Henry Kissinger, and Deng Xiaoping had agreed to former President Gerald Ford's request for a panda film to be made with his son in cooperation with ABC television. But when discussing fund-raising ventures, as it did at every meeting, WWF failed to identify adequately the benefits for China. This showed carelessness, a favored word in the Chinese lexicon.

Charles de Haes then used this meeting to discuss several new WWF fund-raising projects. WWF's aggressive promotion of its interests and its aversion to share public recognition in joint efforts had made conservation organizations elsewhere reluctant to collaborate, and I sensed incipient mistrust among the Chinese too. There would, de Haes explained, be gold coins featuring the panda and other endangered Chinese species. There would also be postage stamps with these animals, he said, as he displayed a sheet with several mockups of the stamps. The bored faces of the Chinese stiffened. "This says Taiwan!" an interpreter exclaimed, jabbing his finger at the Chinese characters on the stamps. Carelessness.

Although four officials from China's Environmental Protection Office had been invited by WWF to Europe the previous June for WWF's annual gala as well as for the ceremonial signing of the protocol and action plan, no one from the two other collaborating institutions, the Ministry of Forestry and Chinese Academy of Sciences, had been invited, a slight with serious consequences. Carelessness. While in Europe, Li Chaobo, the head of the Chinese delegation, was asked to sign not only the agreements on which we had worked jointly but also a new addendum "to clarify agreed commitments" and "to clarify procedures." It was a document prepared by lawyers with their penchant for closing loopholes and demanding certainty where none can exist. Sir Peter, Nancy Nash, and I had argued against an addendum, but to no avail. Those who favored it would have benefited from reading Austin Coates's delightful and insightful book, *Myself a Mandarin,* first published in 1968:

> The Western insistence on obtaining exact facts, making precise definitions and drawing absolute conclusions, breaks down in China somewhere between ninety-five and ninety-eight in the hundred degrees. A Westerner who persists in demanding exactitudes after ninety-five is a bore, after ninety-eight rude. Contracts, oaths, sworn statements, and suchlike are other features of the West which can never be more than 95 per cent respected in China—if that—since they deal with absolutes; and to the Chinese mind, the absolute is essentially absurd, because it does not exist.

Li Chaobo signed the addendum as requested, circumventing standard Chinese procedure, possibly to save face for both WWF and China. But by so doing, he humiliated himself and his delegation. After carefully analyzing the addendum in Beijing, the Chinese found such demands as six-month financial reports including "invoices, receipts" an insult to the dignity of China and an indication of distrust. Li Chaobo had signed without the approval of the collective leadership. On returning to China, the delegation was stripped of watches, photographs, and other WWF gifts. As in ancient times when a messenger bearing bad news was beheaded, interpreter Ma Huanqin, who had translated the addendum, was demoted. The China side refused to ratify the signed agreements.

Charles de Haes apologized for placing Li Chaobo in an "embarrassing position" and withdrew the addendum. It was WWF's turn to lose face.

Qu Geping made a revealing statement during the November meeting in Beijing. "Others said there should not be any kind of cooperation on the giant panda. Mr. Li and I repeatedly persuaded them to cooperate with you." Serious factional strife obviously existed between the Chinese organizations involved in the panda project. WWF's carelessness had provided the pretext for major internal conflict.

The next day's meeting included representatives from all three organizations, the Ministry of Forestry, Academy of Sciences and Environmental

Protection Office, including Zhang Shuzhong, Jin Jiangming, Wang Menghu, and Zhu Jing, the last-named from the Academy of Sciences. Qu Geping's enumeration of problems was repeated. Then Zhu Jing had a surprise for WWF: all equipment that the research center might conceivably need in the future must be decided and agreed upon at once. The Chinese seemed apprehensive that WWF might retract its promises. "My preliminary list is prepared according to work experience in the past," he said. "The list is temporary and incomplete."

I looked at the list with the horrified fascination of someone discovering that the gentle snake in hand is actually venomous. It read like the entire contents of a high-technology equipment catalogue: atomic absorption spectrophotometer, adiabatic bomb calorimeter, electrophoresis, refrigerated centrifuge, sound spectrograph, visible double-beam spectro ultraviolet photometer, on and on. In addition, the list included night cinematic cameras, environmental control rooms, and hot-air balloons. I did not even know what some items were used for, but at least I recognized that the inclusion of two hot-air balloons was ridiculous, of no use in the rugged, fogbound mountains. The list had been hurriedly prepared. Someone had crossed out the original figures and arbitrarily doubled and even tripled the requests for various items. Two environmental control rooms? Most North American universities lacked even one. Yet certain essential pieces of equipment were not on the list.

We had been given no indication that equipment would be discussed this November. When I had left in May, I had been asked to provide details of any items that might require actual modification in the design of the research center. I had brought the information with me, noting special hoods and exhaust ducts for analyses that produce noxious fumes, and I had already given it in writing to one of the Chinese who, I later learned, apparently suppressed it. Since the Chinese had long been deprived of equipment, I could sympathize with their desires, but now greed far exceeded need. Our action plan was adequate, and our understanding of what we wanted to accomplish in the field was, I think, adequate too, but something indefinable in the spirit of cooperation had collapsed, and for that both sides were at fault.

During this meeting, and those that followed on succeeding days, Chinese demands had little flexibility for negotiation. With persistence, if not linear Western logic, they hammered at the problem.

"Principle and detail should be on the same plan; they cannot be separated from one another," said Zhang Shuzhong. "The principle is that the research center will need the necessary equipment. The detail is when will the equipment be delivered?"

"Equipment will be purchased according to need. The need means that equipment is needed for our joint research program," said Wang Menghu.

Once I referred to the research center as a field station—a semantic slip that accurately expressed my opinion of what was needed. It was the wrong buzzword, however, and hours were wasted in assuring the Chinese side that WWF was not abrogating its agreement to build a research center.

On one occasion Zhu Jing said to someone next to him, in Chinese, "Maybe we should delay cooperation with the foreigners." While others were blandly polite, their official smiles strained, Zhu Jing sat with the deep frown and down-turned mouth of an Olmec statue.

Mutual dissatisfaction increased. The project seemed in danger of aborting.

"We have reached a point where there can be no more fruitful discussion on equipment," reiterated Charles de Haes. "George Schaller suggested that foreign experts should be consulted on equipment needed in years to come. There is nothing in the protocol or action plan that says there should be detailed discussion of equipment now; we are not ready to do so now. . . . We cannot accept that we must know every piece of equipment that will be in the laboratory for years to come."

WWF clearly had to take a reserve position: delay decisions on equipment until laboratory biologists could be consulted about what might be needed for biochemical, radio immunoassay, and other work. I did not have the knowledge to help in such matters, nor would I promote the purchase of equipment I considered unnecessary and unsuitable for conditions as they existed in Wolong. It was finally agreed that another meeting would be held in mid-February to discuss equipment.

The ninth and concluding day of our sessions. Before the meeting, Ma Huanqin made a token appearance. She was upset and greeted me with tears in her eyes. Shen Jianguo, a lively and likable interpreter who in days past had worked extremely hard, looked solemn and for no apparent reason took my hand and patted it. What did this portend?

At one point in the meeting, I asked: "Does the November to May program go ahead as planned?" I was concerned about this winter's field work. The action plan specified that the first period of research should begin in November and continue until May. It was now November 24. Would the agreement be honored?

"What is written in the action plan should be valid," replied Wang Menghu, an answer in favor of our original schedule.

Zhu Jing contradicted him, "George Schaller should go back and discuss the equipment with experts."

Zhang Shuzhong expressed a third opinion by handing us a letter that presented the compromise between their political factions: field research may begin, but it will be stopped if WWF fails to submit a list of proposed equipment in January.

"I must have assurance," answered Charles de Haes, "that work between now and May will not be stopped. . . . If you don't give me assurance that we will go on till May we might as well stop now."

Having been constantly on the defensive, we were somehow exhilarated by de Haes's ultimatum. We had been manipulated and maneuvered and we had been made the suppliant in these negotiations while the Chinese were involved in their own bureaucratic games. Now the next step was solely up to the Chinese side.

On November 30, Xiao Shen and I were allowed to leave Beijing by train for Chengdu, and like fugitives we left the city at dawn. The spirit if not the exact schedule of the action plan had been kept; Zhang Shuzhong's opinion had prevailed, at least for now.

Chengdu lies in the Sichuan Basin, an alluvial plain that is sometimes called the Red Basin because of the characteristic color of its soil. It is the richest agricultural area in China and the home of most of the province's one hundred and ten million people. The German geographer von Richthofen found Chengdu in 1870 "among the largest cities of China, and of all the finest and most refined. . . . All teahouses, inns, shops, private dwellings have their walls covered with pictures. . . . No traveller can help being struck with the great artistic perfection of the triumphal arches. . . . In no respect is the refinement more perceptible than in the polished manners and gentle behavior of the people, in regard to which the inhabitants of Ching-tu-fu [Chengdu] are far ahead of the rest of China."

A century has seen certain changes. Chengdu is now a sprawling and polluted city beneath a sky so perpetually gray that, according to a local saying, the dogs bark when the sun appears. For a city with a two-thousand-year history there is remarkably little to interest visitors beyond a few temples, in part because the old city was burned during a peasant uprising in the seventeenth century. The Jinjiang Hotel, where I stayed during visits to Chengdu, is a massive Stalinist edifice. Although Sichuan has one of the world's finest cuisines, the hotel distinguishes itself by serving some of the city's worst food. For those who wish to eat simply, cheaply, and well, there are street vendors and numerous good small restaurants, some specializing in such dishes as *dandanmian* (carrying-pole noodles) or one of many varieties of spiced *doufu* (bean curd). In the eighth century the Tang Dynasty poet Li Bai wrote, "The road to Sichuan is more difficult than the road to Heaven." His words came echoing over the centuries like a prophecy, but I was thankful to have come at least this far.

Not yet having permission to begin work in Wolong, I went to the Chengdu Zoo the day after my arrival to look at the pandas. There were nine of them, housed in an adequate building with an outdoor enclosure. Of the four females in the zoo one had mated naturally and two had been artificially inseminated this past April. One of the latter, a lovely creature named Mei-Mei, had given birth on 20 September to two young (one died) after a gestation period of 157 to 159 days. Pandas have one to two young but they usually care only for the firstborn, leaving any other to die. Of the twenty-three single young or litters born in captivity by 1980, all but one of the births were in Chinese zoos. The exception was the young born to the female Ying-Ying in Mexico City's Chapultepec Zoo. I was eager to get my first look at a baby panda. Taken to a cage screened from the public by grass mats, I was allowed to peer at Mei-Mei and her offspring, a male named Ronshun. She sat in a corner on a bed of hay in a panda's usual slumped posture, with bowed hindlegs out front. Ronshun rested in the hay before her, like an overweight cat in size and shape, at two and a half months barely able to crawl, a cuddly movable toy obviously designed for hugging.

I asked for permission to observe Ronshun a little each day. The zoo directors denied me access to the warm building, where Chinese researchers worked, but they at least allowed me to stand outside in the cold public viewing area. Fascinated, I noted that Mei-Mei groomed Ronshun frequently, licking especially his anal area and ingesting all his urine and feces, thereby keeping the nest clean; and I noted that Ronshun suckled only once or twice in daytime, usually for ten to fifteen minutes at a time. Such details would assist us in interpreting observations in the wild. Standing in raw winter weather, I was warmed by watching the intimacy between Mei-Mei and Ronshun, as illustrated by notes I took one day:

1042. She stands, takes Ronshun's forearm into her mouth, and with the same motion scoops him into her arms and sits, cradling him, biting him lightly in the leg, as

if in play. After licking his anal area, she releases him, and he slides down her sloping abdomen into the hay.
1406. Still lying, Mei pulls Ronshun close and nuzzles his head and chest; he opens his mouth and paws her face; she nuzzles him again, on forearm, neck, and head, and mouths him and paws—not grooming, it seems, but stimulating him—for 11 minutes.

Meanwhile, in Beijing the collective Chinese leadership continued its battle over the future of the panda project. Those factions in favor of adhering to Zhang Shuzhong's compromise triumphed over the obstructionist elements. Three weeks after I had arrived and began waiting in Chengdu, I was finally allowed to proceed to Wolong and begin our collaborative research.

A Footnote to History

I have long been fascinated by pandas. As a high school student, I visited the St. Louis Zoo and watched a panda shuffling around its moated enclosure in the stifling heat. I had read *Trailing the Giant Panda* by the Roosevelt brothers, who on 13 April 1929 were the first Westerners to shoot a panda. Though I avidly read about their quest, I found myself even as a teenager reluctant to applaud their success:

> Our quarry had evidently been in no hurry. For a while he followed up the rocky bed of a torrent, then he climbed a steep slope, through a jig-saw puzzle of windfalls. The fallen logs were slippery with snow and ice. Here we could crawl through and under; there we had to climb laboriously around and over. The bamboo jungle proved a particularly unpleasant form of obstacle course, where many of the feathery tops were weighed down by snow and frozen fast in the ground. Drenched by rain and soaked by snow, whenever a moment's halt was called, we alternatively shivered and panted. . . .
>
> One of the Lolo hunters was close to Mokhta Lone and me. Noiselessly he darted forward. He had not got forty yards before he turned back to eagerly motion us to hurry.
>
> As I gained his side he pointed to a great spruce thirty yards away. The bole was hollowed, and from it emerged the head and forequarters of a beishung [panda]. He looked sleepily from side to side as he sauntered forth. He seemed very large, and like the animal of a dream, for we had given up whatever small hopes we had ever of seeing one. . . .
>
> As soon as Ted came up we fired simultaneously at the outline of the disappearing panda. Both shots took effect. Not knowing where his enemies were, he turned toward us, floundering through the drifted snow that lay in a hollow on our left. He was but five or six feet from Mokhta Lone when we again fired. He fell, but recovered himself and made off through the densely growing bamboos. We knew he was ours. . . .

I was also familiar with *The Lady and the Panda,* by Ruth Harkness, an account of the happy furor created by the infant Su-Lin, who in 1936 was the first panda to reach the West alive. I had searched for and located

Chengdu on the map, for it had been the base of operations of all panda collecting expeditions until the early 1940s.

Years later, in 1963, after I had completed my biological studies at the University of Wisconsin, I remember reading news reports of the first panda birth in captivity at the Beijing Zoo. As a zoologist, I became intrigued by the animal's paradoxes and illogicalities. Science, with its penchant for creating neat categories, had been unable to force the panda into a definite taxonomic position, the animal showing resemblances to both bears and raccoons. Might a study of its behavior help solve the riddle? The panda further stimulated the elusive quality of scientific passion by its peculiar life-style. Here was a bear-like animal devoted wholly to recycling bamboo, a creature as improbable as a carnivorous cow. How had it adapted to such a specialized diet?

My interest, however, was based on more than the challenge of untying biological knots. There was the lure of the remote and rare and the opportunity to be the first Westerner to study the panda's marginal life. As a scientific wanderer imbued with a missionary urge, I also wanted to help the species survive. Research has little fascination and pleasure for me unless an animal also inspires more than intellectual involvement, unless I enjoy its companionship. Like the mountain gorilla, tiger, lion, and other species I had previously studied, an animal must provide an emotional experience if I am to involve myself in its world for years, a perpetual emigrant isolated in an alien culture, afield night and day in whatever weather, as gradually and with care I try not just to know the species but to understand it, slowly unfolding its life like an origami. The panda's real assets are understated and hidden behind its attractive exterior. In short, I thought the panda an animal worth knowing.

Political turmoil, civil war, and Japanese aggression preceded the success of the Communist revolution in 1949. Panda country was closed to outsiders and seemed likely to remain so. But then, in April 1971, the Chinese government sent one of its oblique hints that a policy change was imminent by inviting the U.S. table tennis team to China. I thought it an auspicious time to initiate a panda project. Starting in July 1971, William Conway, general director of the New York Zoological Society, wrote letters on my behalf to the Beijing Zoo, China's Academy of Sciences, and the Chinese embassy in Canada (for China and the United States did not have diplomatic relations then) proposing a collaborative study. He also contacted Sir Peter Scott to make WWF aware of our interests. I was in Pakistan when I received a letter from William Conway: "Peter says that the World Wildlife Fund would be delighted to join in sponsorship of your efforts to study the giant pandas. Peter . . . tells me that neither World Wildlife Fund nor IUCN now has a special project on the panda."

Not surprisingly, the Chinese never replied to our letters, for they were in the midst of the Cultural Revolution, about which we had heard almost nothing. It was surely the major event most poorly reported by the Western press during this century, in monumental contrast to the extensive coverage during and after the Tiananmen uprising of mid-1989. Pandas receded from my mind as I worked on wildlife in the Himalaya, Karakoram, and other ranges. Then, having completed that project, I settled into the swamps of the Mato Grosso in Brazil to study jaguars and their prey. I was in Brazil in the autumn of 1979 when Wayne King, then the director of the Florida State Museum and a person deeply involved in conservation matters, telephoned Kay and left an urgent message for me to call him. A few days later I did.

"George, World Wildlife Fund and China have just agreed on a joint panda study. Are you interested in doing the field work?"

Most certainly I was interested. But being involved in another project, I first needed to decide how to do both. I wanted a few days to think about the matter.

In a letter dated 12 October 1979, to Lee Talbot, then director of conservation for WWF, I gave my answer: "I am tremendously eager and excited about taking part in this project."

Fourteen months later, I joined my Chinese colleagues at Wuyipeng to begin field work. Research aside, I was naturally keen just to meet a panda in its wilderness home, something no foreigner had done for decades. After two hard months of striving for that electric first sighting, the pandas had with uncanny ability still eluded me. At least my failure was not without precedent.

The botanist Ernest Wilson spent many months in panda country, including a visit to Wolong in 1908, but he never saw more than panda spoor:

> It is a solitary animal, and makes beaten tracks through the forest, frequenting the same haunts for long periods, as is evident from the large heaps of its dung which are often met with in the Bamboo jungle.

Most other travelers were no more successful than Wilson in encountering a panda, though some saw captive animals or dead ones.

In 1908, J. W. Brooke hunted in the Wenchuan area near Wolong. His field notes, quoted in a 1909 article of the *Geographical Journal,* noted: "We shot specimens of the takin, serow, goral, parti-coloured bear." The last is another name for the giant panda. This statement referred so casually to an unusual event that one presumes Brooke's local hunters rather than he himself had shot the panda. Killed by Yi tribesmen the same year, Brooke was unable to elaborate on the incident.

A German zoologist, Hugo Weigold, saw a captive panda in 1916, and bought the young animal from a local person, but it soon died.

The first Westerner often credited with the actual sighting of a panda in the wild is J. H. Edgar, who in 1916 claimed to have observed one. But there is serious doubt about that record, as intimated by Brooke Dolan in a letter to Glover Allen, a mammalogist at Harvard, dated 13 February 1941.

> "As for the Reverend Edgar's account of a white bear in a tree between Batang and Derge, I would as well expect to see an Indian rhino on that road. Edgar was essentially a "leg puller" and as far as natural history is concerned did not know the difference between a mouse and a rat. The nearest bamboo is many days travel south of Batang, several hundred miles eastward, the Lord knows what distance north or west. The road from Batang to Derge runs through poplar, prickly oak, barberry, willow, etc. Sowerby, who accepts this yarn, has never been in Szechwan."

The first Westerners who definitely saw pandas alive in the forest were members of the expeditions that went to shoot them: the Roosevelts in 1929; the Dolan expedition in 1931; the Sage expedition in 1934; and a lone British hunter, Captain C. H. Brocklehurst, in 1935. The Dolan expedition, rather than producing another panda obituary, could perhaps have been famous for bringing the first live panda to the West if one of its members, the German naturalist Ernst Schäfer, had restrained himself and captured a six-month-old infant rather than shooting it out of a tree. Instead, Ruth Harkness achieved fame in 1936 for finding a panda cub in the wild and exhibiting it to the world.

That Ruth Harkness, a New York city dweller without wilderness experience, was able to capture a panda is a tale of improbable luck and determination. Her husband, William Harkness, was an adventurer with a private income. In 1934, he had brought the first Komodo dragons to the New York Zoological Society. Later that year, he set out for China to capture giant pandas, leaving Ruth, his wife of two weeks, behind. In Shanghai he teamed up with Floyd Tangier Smith, an American bank clerk turned animal collector. Red armies and red tape prevented the two from reaching panda country. And in February 1936, William Harkness died of cancer.

Within two months, Ruth left for Shanghai to collect her husband's ashes and, as she phrased it, "I inherited an expedition."

In Shanghai, she met two Chinese, the brothers Jack and Quentin Young, whose father had been born in the United States and was an American citizen. Jack, also born in America, had been on the Roosevelt panda expedition of 1929 as naturalist and museum collector. Later, in 1932, he accompanied Terris Moore, Richard Burdsall, and Arthur Emmons to Minya Konka, now known as Gongga Shan, a 24,800-foot peak in Sichuan which the expedition was the first to climb. The two brothers had also collected pheasants, golden monkeys, takin, and other species for museums on an expedition of their own. In spite of their youth—Quentin was only twenty-two-years old when Ruth Harkness reached Shanghai—they were among the most accomplished naturalists working in China. Quentin agreed to accompany Ruth Harkness in part to obtain a panda skin for the Chinese Academy of Sciences in Nanjing. "He wanted to shoot a panda for the glory of China, to show that a Chinese was just as capable as any Westerner," wrote Michael Kiefer in an excellent account of the expedition in the *San Diego Weekly* of 29 November 1990.

After reaching panda country near today's Wolong Reserve, Ruth Harkness and Quentin Young found a panda cub in remarkably short time.

Ruth Harkness described the discovery of that cub in her book *The Lady and the Panda:*

> Quentin stopped so short that I almost fell over him. He listened intently for a split second and then went plowing on so rapidly I couldn't keep up to him. Dimly through the waving wet branches I saw him near a huge rotting tree. I stumbled on blindly, brushing the water from my face and eyes. Then I too stopped, frozen in my tracks. From the old dead tree came a baby's whimper.
>
> I must have been momentarily paralyzed, for I didn't move until Quentin came toward me and held out his arms. There in the palms of his two hands was a squirming baby *beishung* [panda].
>
> Automatically I reached for the tiny thing. The warm furriness in my hands brought reality to something that until then had been fantasy. . . .

Ruth Harkness had trouble with customs when she tried to leave China with the precious cub. She finally sailed from Shanghai with a customs voucher that read "One dog $20.00." She and the cub, named Su-Lin, arrived in San Francisco on 18 December 1936 to begin the international obsession with pandas. Quentin Young remained in the mountains where he ultimately shot two pandas for the Academy of Sciences.

Su-Lin had been destined for the New York Zoological Society, but the zoo refused to buy the animal. Officials thought that the naturally bowed legs and inward pointing toes were caused by rickets. So in January 1937 Su-Lin's odyssey ended at Chicago's Brookfield zoo where she died of pneumonia on 1 April 1938.

But did Ruth Harkness actually find the cub? She was publicly accused of having purchased the animal from a local hunter or from missionaries. And Floyd Tangier Smith claimed that she appropriated his animal, a cub that his hunters found and were saving for him. When I asked Hu Jinchu about the matter, he replied that he once met and talked to Ruth Harkness's guide, a man name Huang Daxin, who told him that she had obtained the cub from villagers.

Speculation about the fortuitous discovery has continued over the years. Of the principals involved, Floyd Tangier Smith died in 1939 and Ruth Harkness in 1947, but Quentin Young lived on, silent on the topic until Michael Kiefer interviewed him at length in California during the late 1980s. Quentin Young's account of the discovery, as he related it to Michael Kiefer, is similar to Ruth Harkness's:

> On November 19, as they were climbing through a snow-covered bamboo forest at 12,000 feet, they heard shots and shouts of *beishung*—panda. Quentin had given the hunters strict orders not to fire if they sighted giant panda. He charged, raging up the hill after them, but when he realized he had left Harkness far behind, he stopped in a clearing and called down to her. Then he heard a baby's cry from a hollow tree. Ruth reached the clearing in time to see him pull a handful of black-and-white fur from the upside-down V of the panda den. It weighed less than three pounds and had not yet opened its eyes. . . . The hunters had sighted an adult, presumably the cub's mother, and claimed the adult had escaped them. But Quentin believes they killed it and ate it and sold the pelt."

Did Quentin Young show flexibility in reconstructing his memories?

"If I lie, do I gain anything?" he answered when queried.

Jack Young shot a panda in 1937, probably the last sighting of a panda in the wild by a foreigner. I hoped to be the next to see one.

The pandas, however, were extraordinarily elusive at Wuyipeng and during weeks of tracking them in late 1980 and early 1981 I never encountered one in the snowbound forest. Late one day, unseen by any of us, a panda had even travelled past our tents, and that night passed the tents again, as if taunting us with its tracks.

2 March 1981. At dawn I leave for Bai Ai where Xiao Wang, on his way back to Wuyipeng after checking traps yesterday evening, met a panda ambling along the trail. A light snow has fallen during the night, and within minutes of camp I find fresh tracks going uphill. Slowly I follow the prints, placing my feet softly to avoid snapping branches beneath the snow, my senses straining to perceive any motion, any sound. Three droppings are still warm to my touch. Ahead a twig breaks, then silence. I move several steps. Bamboo rustles, and once more a tense hush. The panda is obviously trying to detect me too. Unseen we confront each other, the atmosphere almost tangible with emotional intensity. But having perceived potential danger, the panda withdraws, and, disappointed, I retrace my route.

Nearby is yet another fresh track, that of a smaller animal, probably a subadult, and it is traveling parallel to the one I have followed. Howard Quigley decides to find out where the small animal has come from, while I backtrack the other. Around dawn the larger panda had approached to within one hundred feet of our tent, detected something worrisome, retreated a little, then arced around camp on its way uphill. I continue downslope, following its tracks through dense umbrella bamboo, an impenetrable tangle of stems, their snow-laden tops bent into a solid canopy. Unable to bend the stout stems apart, I crawl through them, head facing downhill, smashing dead stems aside with stiff, raw hands. Near the base of the slope, I come upon the panda's bed at the base of a maple.

On my way back to camp, I meet Howard, his clothes as sodden as mine, dead bamboo leaves clinging to his hair and beard. His panda had also gone down into the valley. Did the two travel near each other by accident? I want to follow the animals uphill, but fog has already erased the upper part of the Choushuigou and it might soon snow again, obliterating all tracks. As the pandas have angled uphill, they might cross a trail of ours above Bai Ai, so rather than following the animals, I will try to bisect their tracks there. First, however, I go to the tent where Kay is nursing a fire in our obstinate stove. Crowding the stove until my clothes steam, I sip the hot tea she immediately hands me.

The trail above Bai Ai follows the rim of a valley. As I ascend, a moaning hoot fills the silence, soft yet resonant. Ahead I see the still figure of Hu Jinchu, and when he notices me he points to a tall, lone spruce downhill of us. There, crouched on a bough near the top, is a small panda. "Finally," I say to myself. At 5:15 P.M. it hoots again, sending its plaintive call across the hills. A few minutes later it calls once more, and as the sound fades we spot another panda, a bulky adult, among the dense lower branches of the same tree. It descends ponderously, hindlegs first, sliding the last feet in a shower of bark, and disappears in the bamboo. We think it is a male who for reasons known only to himself is ill disposed toward the other. Relieved of its tormentor, the small panda gives up its insecure perch to huddle against the trunk of the spruce. In determined disregard of the cold and gathering darkness it quietly remains there, a lustrous fragment of life, idealized but real, as fog fills the valley, linking the slopes, arriving so quietly that we are aware of its coming only when the panda fades from view. Hu Jinchu and I smile at each other. It is a day of double happiness.

5

A Mountain of Treasure

March to June 1981

In early March, the snow had turned slushy, retreating up the slopes in a first intimation of spring. Yet the traps we continued to check, with greater urgency each day, remained empty, despite our intense hope of catching at least one panda before season's end. Pandas had come tantalizingly close to traps on several occasions. Once a subadult headed directly toward a foot snare, its tracks filling me with joyful anticipation, but too soon they veered away and down a steep slope. Whenever the animal came to a forest opening where snow was deep and bamboo sparse, it tobogganed downhill on chest and belly, leaving a deep furrow, and I thought how much I would have liked to witness its lonely winter sport. Another time a panda circled a log trap, attracted by the smell of meat within, but could not enter because a weasel had earlier tripped the trigger, closing the door.

By 9 March, knowing that he had only one month before leaving, Howard lamented: "This is the first day I feel discouraged."

10 March. The next morning while checking traps, I spot Wang Lianke, or Xiao Wang as we call him, hurrying in my direction along the trail. Xiao Wang's slight build, high cheekbones, and upturned nose give him the appearance of a youthful forest spirit, at home among mossy glades. He is the best of our field staff, steady and dependable. Now, his eyes shining with excitement, he is looking for Howard or me. A panda has been caught near Bai Ai. He encircles one wrist with his hand to indicate a snare.

We soon find Howard. "Are you ready to collar a panda?" I ask.

"Well," he drawls, smiling broadly.

As everyone in camp hurries in single file along the trail, I am elated, and also worried, my stomach knotting with tension. To radio collar the panda, I will have to immobilize it with a sedative. And immobilization is always dangerous because an animal may react unpredictably to the drug. An allergic reaction or any number of mishaps may kill it. True, I have been well briefed by New York Zoological Society's veterinarians, and we have a drug—Telazol, also known as CI 744—that is especially well suited to carnivores. However, we will be working on not just any animal, but on a rare one, a national treasure, a symbol whose accidental death would forever haunt us.

The panda sits crouched at the base of a tree, one forepaw held by the snare. It has clawed the trunk in a brave and lonely effort to free itself, and now it waits in the twilight of rhododendrons, a gentle creature with puzzled eyes looking into an uncertain future. We gather silently about thirty feet from the animal, a small one.

"How much do you think it weighs?" I ask Howard in a whisper. Drug dosage depends on an animal's weight.

"At most eighty pounds, judging by black bear," he replies.

Howard prepares a syringe and mounts it on the tip of a six-foot jab stick designed to administer a sedative to a wild animal from a safe distance. Slowly he approaches the panda, which waits without sound or show of defense. As Howard comes near, it moves to the other side of the tree, and it remains quiet there even when the needle jabs it in the shoulder muscles.

"Got it!" Howard exclaims. But on examining the syringe he finds that the needle is bent and only a little of the drug has been injected. "Its hide sure is tough," he comments, and tries again, this time successfully.

But the panda does not sleep deeply until it has received a third dose. With the animal finally quiet and on its side breathing steadily, we work quickly, uncertain when it will awaken. I check its paw. Trapping has not hurt it, I note with relief; the snare has held without abrading the wrist. Measurements: tip of nose to tip of tail 138 cm (54 inches), shoulder height 71 cm, tail length 8 cm. Hu Jinchu checks the animal's sex. A male. The testes are easy to palpate in the lower abdomen; they are not yet in the scrotum, as they would be in an adult. The panda is about two-and-a-half years old. Perhaps we have met earlier this month, on that foggy evening when for the first time I saw a panda. If so, his is a hard-luck story; he has been chased to the top of a spruce by another panda and trapped by us, all within a couple of weeks. We tie a spring scale to a pole. Then to weigh the panda, two of us lift the pole high and raise him suspended by his legs: one hundred and twenty pounds. No wonder another dose has been required to sedate him; we have greatly underestimated his weight. We fit him with the radio collar, not tight because his neck will grow, but not so loose that he can slip the collar over his head. Soon he breathes more heavily and begins to stir. We cover him with a net so that he will not stumble away half-drugged and fall down a cliff, poke a stick into his eye, or otherwise injure himself.

"When will you tune in on him?" Kay asks casually, wondering when we will check the radio signal.

Howard and I look at each other. I ask, "Have you taken off the magnet? I haven't." Howard quickly fumbles under the net and pulls it off. To activate the transmitter this magnet must be removed. In our excitement, we have almost released the animal with a useless radio.

"Don't tell," he says with a wry grin.

We watch the panda struggle from the net and hurry down-slope. I check the time: 11:15 A.M. A magic moment as our first collared panda vanishes into the protection of the bamboo. The release of tension is palpable as we shake hands all around. I now realize how damp and cold it is and that my clothes are too thin, and that the sun's feeble rays still bespeak winter. But on a slope, oblivious to patches of snow, I find a lavender primrose, the season's first, a speaking-of-spring flower, as the Chinese call it. With buoyant steps we return to camp. Meanwhile the panda moves steadily uphill and over the ridge into the Zhuanjinggou. Now in all that expanse of bamboo I will know where at least one panda leads its solitary existence.

Technology potentially increases rather than reduces work, as our camp soon discovered. With a radio transmitter now sending signals for nearly two years until its batteries are exhausted, we had an opportunity to locate an animal and monitor its activity continuously both day and night. We decided to determine the location of the animal once a day and record his activity for five whole days every month to discover if seasonal patterns exist.

12 March. Two days after the collaring, Howard and I are on the ridge above camp to pinpoint the panda's position from the numbered triangulation points—4X, 11X, and others—we have established there. Deciding then that it will be useful to monitor the animal continuously for twenty-four hours, I return to camp while Howard takes the daytime shift.

In mid-afternoon, as Kay helps me pack food and equipment for a night in the forest, I hear a call, and looking out of the tent I see Zhou Shoude running toward me, waving his arms in a spontaneous and unguarded reaction, shouting "*Daxiongmao!* Panda!" Lao Peng has just returned and reported a panda in a trap located along the rim of the Zhuanjinggou. Lao Peng, Zhou Shoude, Xiao Wang, and I hurry up the mountain, carrying sedation equipment, sleeping bags, and other items, moving so rapidly that sweat soon soaks us in spite of the raw weather.

The panda almost fills the trap—a converted animal transport cage of aluminum reinforced with logs for stability. She gazes at us and gives a nasal honk, a pathetic sound of distress. When we draw closer, she snorts softly in mild threat, and then gives a peculiar chomping sound made by rapidly opening and closing the mouth, teeth clicking and lips smacking, to show her anxiety. When this does not deter us, she emits a gorilla-like roar so explosive and loud that we involuntarily turn to flee. With nervous laughs we look at her again, a female we would later know as Zhen-Zhen, an animal that became more memorable than any other, one that in many ways came to symbolize the project.

It is too late in the day to collar her. I ask Xiao Wang to fetch Howard. Both of us will spend the night here to ensure the female's safety and to monitor the other panda. There is no place for a comfortable sleeping site in this thicket, but at least we each find a place to lie down. Howard takes the first shift, listening every fifteen minutes to the panda's signal. I rest fitfully, stifled by bamboo, uncomfortable with a rock jutting into my sleeping bag, and tense with worry. This female chews logs from the trap into chips with exceptional diligence. What if she should choke on a piece of wood? I can hear her breathing heavily, and with the beam of my flashlight I seek the white reflections of her eyes. She is all right. As night dew settles on my face, I doze until it is time for me to take over the monitoring.

We arise at dawn to await the arrival of the others. The collared panda has been inactive for hours, and now as 7:30, 7:45, 8:00, 8:15, and 8:30 pass without change in signal we become concerned. Surely a panda would not sleep so far into the day. Has he somehow removed his collar?

With twelve of us, including all nine Chinese from camp, the small clearing around the trap is crowded and noisy, and the panda roars and swats the air in anger. I shoo everyone back from the trap and explain yet again that an animal must be tranquil during sedation, especially when going to sleep and waking up, to prevent injury when its actions are uncoordinated. Howard injects the drug. Immediately most of the Chinese crowd around the trap, talking loudly as they photograph and look, and the panda becomes frantic. My pleas are ignored until I say fiercely, "Get back!" Xiao

Xie, a student interpreter whose English is rudimentary, translates this as "Get out!" which naturally ruffles feelings (and the incident was promptly relayed to Beijing and from Beijing to WWF).

9:37 A.M. The panda's head sinks slowly until her nose almost touches the ground. When I reach into the trap to shake her gently, she does not respond. We remove the door, then pull her out and stretch her on a tarp. Like a surgical team in an emergency ward, each of us concentrates on a task. I smear salve on her unblinking eyes to prevent their drying out. Pan Wenshi, who has just returned for a one-month tour of duty, searches the dense fur for fleas and ticks without success. Hu Jinchu squeezes the panda's flaccid nipples but finds no milk; she does not have a young now, but may have lost one earlier in the year. Howard takes her pulse—sixty-eight beats per minute—then measures her, calling out the figures for me to write down: total length 166 cm (65 inches), shoulder height 81 cm, tail length 13 cm. We pull her mouth open. Her yellowish canines show the bluntness of middle age, but we unfortunately have no reliable means of determining her exact age. The fourth premolar and first and second molar on the left side of the lower jaw are missing, and a thick layer of tartar, gray and hard like limestone, encrusts several teeth. Does this affect feeding efficiency? Is this why she is lean, pelvis sharp beneath her hide? It takes a communal effort to weigh her: 190 pounds. Having fastened the radio collar around her neck, we now wait for her to recover.

10:35 A.M. Her eyelids flicker and suddenly she barks and rakes the air with a paw. Gingerly we pull her back into the trap where she soon raises her head and focuses her wistful eyes on us.

11:45 A.M. She sits fully coordinated in the trap, greeting with roars our effort to raise the door. I expect her to bolt for freedom. No response. For twenty-five minutes she faces the door-less entrance before tentatively peering out; then, as if recharged with energy, she suddenly rushes into the bamboo where agitated stems briefly trace her progress.

I feel drained, limp. Even when Howard tells me that the other panda has finally become active at 10:15, his radio obviously functioning, I can do little more than show subdued elation.

After dinner the next day, we had a communal meeting to discuss possible names for the two collared pandas. It was not aesthetically pleasing to refer to them by their distinctive radio frequencies, 188 for the male and 194 for the female. Giving a name to a panda is a weighty matter, not something to be done casually or carelessly, for with a name the animal ceases to be just one among others in the forest and instead becomes a member of our community, one whose behavior we will observe, debate, and judge as much as our own. A name must be appropriate. Hu Jinchu suggested the name Long-Long for the small male, a fine choice with which we all immediately agreed. "Long," meaning dragon, honors the name of the reserve. And it is also apt in another way. Every dragon has nine sons, each with his own characteristics—smoke, river, fantasy, food, battle, and others; one of these sons is a recluse like the panda, as symbolized by a dragon painted on closed doors in China. Agreement on a suitable name for the female was more difficult to reach. Various names were suggested and discarded because they were unsuitable, had been used for zoo animals, or for other reasons. Finally Pan Wenshi offered Zhen-Zhen, meaning "rare," "precious," "treasure," and "pearl," all fitting tributes, and in this manner Zhen-Zhen began her new existence.

During subsequent days Long remained in the Zhuanjinggou, whereas Zhen roamed on both sides of the ridge as far downhill as Erdaoping, where we also had a log trap. A week after being collared, she entered that trap; released, she meandered back over the ridge and into the trap in which we had first caught her. There she looked at me with tense irritability, and when I opened the door she immediately lumbered off. The following day she sat in the Erdaoping trap again. She obviously knew a good thing when she smelled it and appeared to have made a trade-off: a meal in exchange for a bit of stress and inconvenience. While I admired her speed in adapting to new circumstances, I viewed her pragmatism with disfavor because it might well prevent us from catching other pandas in those traps.

17 March. At noon I climb toward the ridge carrying bedding and equipment to spend the night monitoring the activity of Long and Zhen. The day can not make up its mind whether to be winter or spring, the sun unable to struggle out from behind the clouds for more than a minute. The ground steams softly, pungent with the odor of moss and moldering leaves, and wolf spiders dash between boulders wet with melting snow. Rhododendron leaves have unfurled. A nuthatch in a birch grove gives a long trill in a downward crescendo like a girl descending a hillside with light and happy steps. Slowly I climb back into winter where snow is still deep in patches and tree buds are not yet swollen. On a knoll, at triangulation point 11X, I sit down beneath a large fir and begin monitoring the two pandas. Below me arrow bamboo slopes away until stopped by a wall of conifers. With about a third of their leaves dying in winter but still clinging to their branches, the bamboo has the yellow tinge of a grass meadow in autumn. I snack on peanuts and crackers. A striped squirrel, resembling a chipmunk, dashes between trees. When I flip it a piece of cracker, it scrambles up the tree, where clinging upside down on the trunk, it watches me with glistening eyes. Two brown-crested tits flit past. They seem to be paired, unlike their winter custom when this species often travels in mixed flocks with tit-babblers, coal tits, and white-eyes.

The breath from the slopes above becomes colder and the bamboo leaves and I shiver. I inflate the air mattress and unroll my sleeping bag, protecting these and my pack beneath a tarp tied to surrounding bamboo and a viburnum. It begins to snow, the flakes large as cherry blossoms. Tucked into my sleeping bag, I look at the dark firs looming into whiteness; as fog drifts in, the trunks become gray blurs, shrunken and colorless, and finally dissolve as dusk turns to darkness. Snow scurries before a breeze and settles on my face. Aloof on my narrow ridge, I make contact with the pandas. I tune the receiver to Long; he is below me in a ravine, feeding in spite of the bitter night. I switch the dial to Zhen; she is in her usual haunts farther along the same slope, her signal calm and constant. I can imagine her there, curled up against the elements, flakes settling on the black and white of her coat until she is no different than the snow and the shadows between the trees.

Lying there either dozing or looking at the stillness and sometimes at the glowing sliver of moon, I wait for the minutes to pass until it is time to make contact with the pandas once again. When the slopes and trees begin to detach themselves from the night, my world is white, clouds so merged with snow that I can barely distinguish one from the other. But soon there is sun and the fir boughs drip as if in a rainstorm, clouds descend and fill the valley, cool like a glacier, a river of light so smooth it seems I could ski over it. Above, the ridges are sharp-edged against a sky flushed with pink.

At 9:30 in the morning Hu Jinchu and Pan Wenshi come to relieve me. I am ambivalent about descending the ridge. Solitude does not necessarily provide tranquility, but the long, solitary hours focused on the pandas, detached from day-to-day events and problems, remain a temptation. Here thoughts can be given unbounded to the night's silence, here one seems whole. Below, the day cannot begin and end with my mind serene and totally devoted to pandas. There I am divided. I am an administrator worrying about an April meeting in Chengdu, about a late March artificial insemination program at Yingxionggou, and about a scheduled succession of photographers and other intrusions on Wuyipeng; I am an expatriate lost among strangers in a strange culture; I am a biologist under pressure to produce results that will enable two bureaucracies to justify the expense of this project; I am a conservationist watchful of the different kinds of menace threatening the panda. But now I also want to return to the tent, to our home, as any place becomes a home where Kay and I expect to stay a long time, and I want to tell Kay of my night and hear of her adventures. I descend. Kay tells me of a brown weasel entering the tent in search of mice, and of Qiang women visiting to look and chatter and be photographed with our Polaroid camera after having carried baskets of cabbage and onions and other supplies to the camp kitchen. Then we do the many small tasks demanded of a home, such as carrying in firewood and cleaning out ashes. On this day, as on others, there are letters to answer or reports to write or bamboo samples and panda droppings to dry.

We lived in one world, the Chinese in another, and we coexisted mainly at meal times and on the trail. Not that this was our choice. I have never tried to make the ambiguous effort to adapt totally to another culture, trying to act like a native, except to modify thought and attitude where courtesy demands it. I do not have the romantic notion that in another culture I can ever cease being a stranger at the periphery and become an insider. But I did hope when beginning work at Wuyipeng that acquaintance might lead to true companionship, a sharing of ideas, thoughts, and feelings, as it had in other countries. Of course I had long been aware that Chinese are considered reticent, that they build social fences around themselves. And I further realized that in a situation such as that at Wuyipeng my coworkers might at first respond to their own image of another culture, looking upon me as the foreigner or American rather than as an individual. With mutual respect and attempts at understanding, I assumed trust would soon come. I was wrong.

In China, a "policy of opening to the outside world" was now in effect but, other factors aside, the Cultural Revolution was still so much in everyone's mind that it continued to influence most actions. It was not just a Gang of Four but a large segment of society that had been gripped by an epidemic of psychic madness to eradicate the Four Olds—ideas, culture, customs, and habits. Yet without the old the new is meaningless. It was a time when the logic of "class struggle" was taken to an ultimate absurdity. One hundred million people were labeled reactionary and "forced to eat bitterness." All intellectuals were automatically placed in the "stinking ninth category" following landlords, enemy agents, capitalist-roaders, bad elements, and others, categories so sweeping that even having a relative abroad was a basis for persecution. Since accusations brought rewards and failure to report an offense was a crime, everyone kept close watch on each other either to inform or in fear of becoming a target. The worst in human nature was exploited, and ultimately the aim seemed to be not to find guilt but to attack the accused. Like the Spanish Inquisition of the sixteenth century, the Cultural Revolution practiced sheer terror and also made this terror a pretext for social gatherings and processions. Heretics, now called counter-revolutionaries, were publicly forced to recant sins they had not committed, and they were criticized, ridiculed, attacked, maimed, imprisoned, tortured, and murdered by a general public that either participated or stood idly by. So many hopes were ruined, so many lives shattered that the nation's collective fear, and perhaps its guilt and shame over these "misguided" years, will persist like a wound on the land until the last survivors have joined their ancestors in the second half of the twenty-first century.

In our camp both victims and possibly tormentors lived smilingly side by side, watching each other, still keeping mental if not written dossiers, uncertain how soon history might repeat itself. Each department or unit retained its Public Security Bureau whose task was to supervise everyone's actions, giving permission to marry, travel, change jobs, have a baby, invite a foreigner home, everything. No one dared visit our tent alone, no one wanted to seem overly friendly to a foreigner; fear of denunciation pervaded the camp. Everyone was his neighbor's policeman. As a Chinese told me: "I must be fearful of everything I do or say." No wonder everyone escaped the social claustrophobia of our camp as much as possible. Though capable and intelligent, most Chinese with whom I worked had had their imagination and initiative stifled, they had to become willows to survive—passive, afraid to take risks or excel; they had to live by the lesson of the Chinese proverb: "Tall trees are cut down." They lived in a society where bending is a virtue, resistance or determination a crime. True, the Chinese as a people were infinitely better off than at any time in their history. Their health and education had been vastly improved and everyone had been raised to more or less an equal standard. But they deserved much, much more. Kay and I felt a great empathy for our colleagues and sorrow for ourselves because we dared not express more than restrained friendliness to anyone for fear of exposing him or her to criticism, or worse.

Evenings, I sometimes gave an informal talk to our group, discussing the meaning of the data we had collected and presenting overviews of such topics as territoriality and conservation. I wanted to strike a responsive spark and provide an answer to the dispirited rhetorical question posed by one researcher, "What's so special about the panda?" But I liked best those evenings when Hu Jinchu told us about pandas in ancient times and about the people of Sichuan and their customs. A Confucian gentleman, Hu Jinchu was self-deprecating and private, his feelings masked by a laugh. We heard he had suffered tragedy during the Cultural Revolution but were never told the details, and we did not pry. He had studied Yangtze River sturgeon until he was ordered to do panda research. Mornings he practiced *taiji,* the slow motions of ritual combat, behind the tent. I was fond of him, but given the circumstances, we could only form an alliance, not a friendship. Having begun to read his face, I saw that his laugh was mirthless, seldom involving his eyes, yet when he talked of history, his favorite topic, he sparkled with good humor.

Knowing that in China everything from tiger clavicles and bear gall bladders to flying-squirrel droppings are used in traditional medicines, I asked him if the local tribal people use pandas in any such way. Only urine is said to have medicinal properties, he told us, and for but one purpose: to dissolve a swallowed needle. At least demand is unlikely to be very high. Before Liberation in 1949, he continued, pandas were much hunted for their hide. To sleep on a panda pelt confers two benefits: it keeps away ghosts and it foretells the future, in that if the night was comfortable all will be well. However, long ago during the Qin Dynasty in 220 B.C., the earliest dictionary, the *Er Ya,* described another medicinal use: the panda's pelt was useful in controlling menses. The *mo,* as the panda was then called, was also said to eat copper and iron. Other ancient books, too, mention such a metal-eating habit. For example, *The Classics of Seas and Mountains,* a famous geography text dating from 770 to 256 B.C., mentioned that "a bearlike, black-and-white animal that eats copper and iron lives in the Qionglai Mountains south of Yandao County." And *Shuo Wen,* a dictionary of Chinese characters from about 100 A.D., noted that the *mo* "has a whitish color, and resembles a bear. The animal licks iron and may eat ten jin [ten pounds] at a time; its pelt is good for keeping warm." Hu Jinchu believes that the supposed propensity of the *mo* to eat metal may be based on the fact that pandas sometimes enter villages and lick and chew up cooking pots. A panda certainly has no problem crunching today's aluminum pots in its powerful jaws. One animal at Yingxionggou shredded its water pan and hours later defecated a handful of aluminum nuggets.

In ancient China the panda was already considered rare and precious, Hu Jinchu told us. When Dowager Empress Bo, mother of Wen Di, fourth

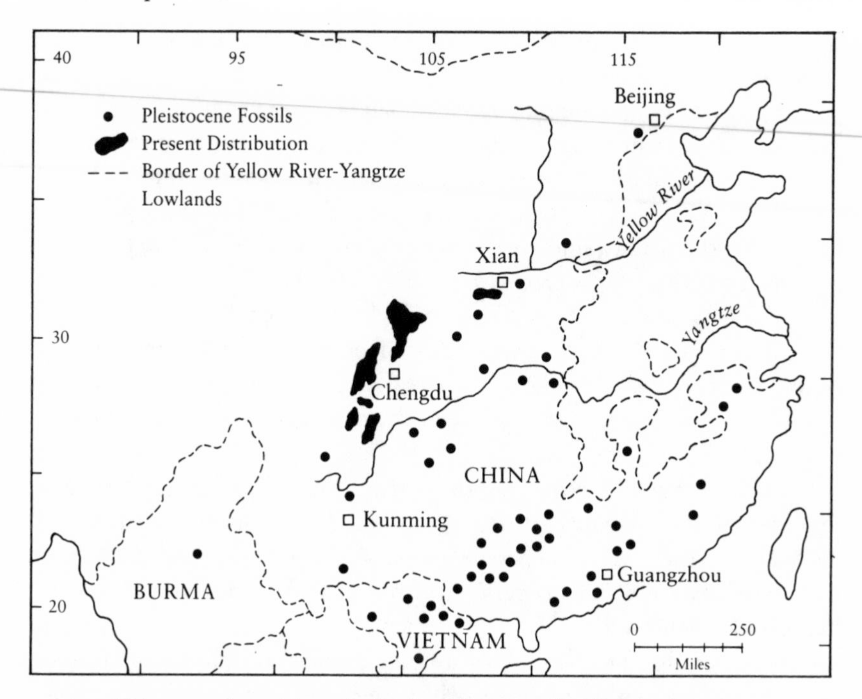

Pleistocene sites with panda fossils and present distribution of pandas

emperor of the Western Han Dynasty, died and was buried in the Nangling mausoleum near Xian in about 170 B.C., the skull of a panda was entombed with her. The grandson of Tang Taicong, first emperor of the Tang Dynasty (618 to 907 A.D.), may have sent two live pandas, as well as skins, to Japan as a token of friendship, a political gesture not repeated until recent decades.

During the Western Han Dynasty, which lasted from 206 B.C. to 24 A.D., the emperor's garden in Xian held over forty rare animal species, according to the noted historian of that period, Simao Xiangru, and the most treasured was a panda. Not until the 1950s were pandas again on exhibit in China's zoos.

On 11 March 1869, a hunter brought a panda skin to the French Jesuit missionary Pére Armand David, who then made the panda known to Western science. My perspective on the panda had been limited to the hundred and ten years or so since that discovery. Talking with Hu Jinchu extended my horizon by over two thousand five hundred years.

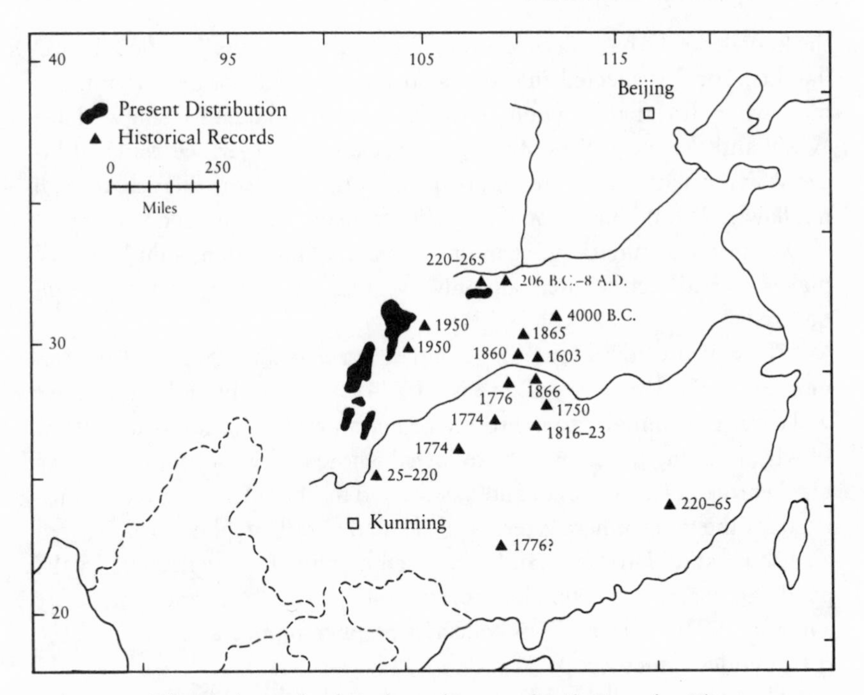

Historical records of pandas and present distribution of pandas

In addition to field work, the panda project included a program to improve reproduction in captives. Emil Dolensek, chief veterinarian at the New York Zoological Society, and Stephen Seager, a reproductive physiologist at the National Institutes of Health, would soon arrive to spend a month collaborating in an artificial insemination program at Yingxionggou and the Chengdu Zoo. With both men due in a few days, I went to Yingxionggou to check on working conditions. As I approached the station, I could hear an animal screaming. One man held a goat that was rigid with shock while Bing Ji, the chicken veterinarian, poked inside its rectum with an object that looked like an electrified corncob. It was a rectal probe used in artificial insemination, and Bing Ji was practicing his skill at electroejaculation before applying it to pandas. The goat later died. I was told Bing Ji had injected a panda with a hormone to induce estrus, giving three times the required dose. It was possible that he had disrupted the panda's reproductive cycle just as our collaborative work was to begin. His reference book was an American endocrinology text from the 1930s; at least the knowledge he had was not wrong fifty years ago. Bing Ji radiated unreliability, and I suspected that our association might not be smooth. My main worry was that he would harm the pandas and then somehow blame WWF and the New York Zoological Society. However, we watched his every action with such care and dedication that our surveillance, though not flawless, could have won high praise from any security department.

Work on insemination began on 28 March. Shan-Shan, a middle-aged male who had been wild-caught only a year before, was the first to be tranquilized.

"We will use this blow gun. . . . The syringe was developed in Germany. . . . The drug we will use is CI 744. . . . After immobilization we will carry the animal into the room for insemination," Emil would intone slowly, outlining the steps to be followed before each procedure. He was a big, bearded, bear of a man and a dedicated teacher. An interpreter translated for the two Chinese veterinarians and the staff. Stephen then inserted an electric stimulator in Shan-Shan's rectum while Emil held a cup by the panda's penis, which is small, purple, and shaped like a sea anemone. Only a minute 0.9 ml of semen was collected. Stephen immediately checked the sperm under a microscope.

"The sperm are all dead. Dead," he lamented. After adding a little warm saline solution, he noted some life: "Perhaps five percent motility. It should be eighty percent."

Insemination would not be successful with such a low sperm count. This made today a practice session. The female Li-Li had been caught in the wild two years before. She had just finished her period of heat before Emil and Stephen arrived; for unknown reasons she had not been given access to a male while she was in heat. After she was tranquilized, Stephen inserted a laparascope—a tube with a light for examining the uterus—and confirmed that her tissues were neither red nor swollen as one would expect of an estrous female. He then pushed a plastic tube up the laparoscope and injected the semen, completing the insemination. (In defense of Shan-Shan, I should say that two days later he produced 9.0 ml of semen with a seventy-five percent motility, and these sperm were used to inseminate Li-Li again as well as another female, but neither conceived. However, Stephen was toasted at a going-away banquet for making Shan-Shan and other males happy.)

Most panda infants in zoos have resulted from artificial insemination, not by choice but because captive animals rarely breed naturally. Although pandas have been maintained in zoos for half a century, we still know too little about the correct way to manage the species. Should animals be kept separate and brought together only when the female is in heat, a method successfully used on cheetahs, also a solitary species that once failed to reproduce in zoos, or should they be kept in pairs, or in groups? All these methods have been tried, none with consistent success. It is known that females usually come into estrus once a year between mid-March and mid-May, though occasionally they come into estrus at other seasons too. Their receptivity lasts for about twelve to twenty-five days, and peak receptivity is limited usually to two to five days. During this period the behavior of a female changes; she loses her appetite and becomes restless, her vulva swells and she rubs it against objects. One captive female I observed backed up to a male, soliciting; she rolled on her back in front of him squirming and writhing; she reached for him gently with her forepaws, trying to pull him toward her, her invitation explicit, her desire obvious. But he merely watched her passively. A male that was sent from the London Zoo to the National Zoo in Washington, D.C., for a romantic tryst attacked and seriously mauled the estrous female. It appears that males, too, must come into heat before showing an interest in females; indeed testes-size changes seasonally, being largest in spring. Male and female must synchronize their courtship, a harmony that for unknown reasons captive pairs usually fail to achieve.

Emil, who had a deep concern for animals, worked with the Yingxionggou staff to prepare a detailed management plan that included details on feeding, cage cleaning, and so forth. He had shipped over a crate of expensive veterinary supplies and he patiently explained the use of each item, making certain that dosages, techniques of administering drugs, and other aspects were carefully written down. Stephen designed a chart for recording behavior and physical signs of heat in females so that next year the animals would be inseminated during the best possible time. Emil and Stephen both did their best. In a report to the Chinese, Emil stated mildly:

> A two week observation of keeping practices . . . revealed that more sensitivity to the animals' needs is required by the keepers. . . . It is important that the keepers should develop a rapport with their charges. . . . The animals at Yi Shoung Gou have responded to this insensitivity by remaining relatively wild and unresponsive to the keepers.

In May, I checked to see if Emil's visit had benefited the pandas. The management plan was ignored and the veterinary supplies were unused (nor would they ever be used), even the floor disinfectant. Most pandas have roundworms—a species, *Ascaris schroederi,* unique to pandas—and all captives there had them; one female had just passed a writhing handful of one hundred large worms. The staff used only herbal medicines, instead of the more efficient medicine we supplied, and the animals constantly reinfected themselves in the dirty cages by eggs that passed in their feces. Emil made another valiant effort to improve conditions a year later, but soon there was neglect again, not out of deliberate thoughtlessness but from lethargy. As someone there told me, "I don't care what happens at Yingxionggou as long as the pandas stay alive and I don't get into trouble."

Another time I went to Yingxionggou to accompany Lars Eric Lindblad and Nigel Sitwell, who had come to Wolong to assess its tourist potential. The one surviving panda youngster, then two years old, was permitted into the outdoor enclosure because we were there. Released from its dark cell, it exploded with joy. Exuberantly it trotted up an incline with a high-stepping, lively gait, bashing down any bamboo in its path, then turned and somersaulted down, an ecstatic black and white ball rolling over and over; then it raced back up to repeat the descent, and again. It gave me a glimpse behind the panda's tranquil facade, it emphasized the imagination there.

With fervor that owed more to anguish than to optimism, I again told Wang Menghu, who was in Wolong at the time, about conditions in Yingxionggou, noting that some of China's zoos treat their pandas quite well, and that Wolong should do no less. To have keepers without interest was no longer enough. He agreed and noted that his attempts to make changes had been less than successful, that Beijing in this and other matters finds it difficult to deal with local staff, for, as a proverb says, "A dragon is no match for a local snake."

12 April. Around-the-clock monitoring of Zhen and Long began on 10 April, and today it is Kay's turn and mine to take the night shift. A small, orange-colored tent has been erected on the ridge at triangulation point 4X. Zhen is in a ravine near Erdaoping, Long as usual is in the Zhuanjinggou. The afternoon is damp and chilly, and we are soon in our sleeping bags, antenna mounted on a pole, the receiver between us, ready for the night's vigil. At 5:45 in the afternoon we hear a panda call a few hundred yards away, down by Erdaoping, a medley of moans, hoots, yips, and growly barks, a barnyard of animal noises. I dress hurriedly, descend to the area from which the sounds have come, and wait there for an hour in the cold stillness until a soft moan seeps through the bamboo. With loud panting and the scratching of claws on bark, a panda suddenly clambers up a nearby fir. His back is broad and muscular, his movements are powerful, and the silver of his coat is striking in this somber forest. He is a formidable creature, reminding me of the male mountain gorillas I had studied two decades before among the mist-shrouded peaks of central Africa. He reaches the lowest branch twenty feet up and settles himself there as if to send his calls across the valley again. However, he descends after just fifteen seconds, possibly because he has seen me. About a quarter mile away, Zhen certainly hears his insistent calls, which continue at intervals until almost midnight, yet she is not swayed by his amorous song and stays where she is.

13 April. Kay and I return to camp after Xiao Wang takes our place at 4X. But I climb the slope again, hoping to tape-record the panda's calls. I have barely reached the site of my evening encounter when a roar shatters the forest. It is soon followed by the yips, moans, and barks of two animals. Zhen has joined the male. Both are invisible beneath a blanket of bamboo in a small, shallow valley. At 3:55 that afternoon, I spot the head and back of the male angling toward a rivulet, where he pauses to drink. Downhill from him, a third panda hoots softly and the male answers similarly, hoo hoo hoo, the two calling back and forth for a minute, each expressing uneasiness. Then the male moves abruptly back toward Zhen, who has been patiently waiting nearby, bleating like a nervous goat to indicate her friendliness. And following the male, tagging thirty feet behind, is a second male, a newcomer smaller than the first. The big male wheels and charges, crashing through bamboo; the smaller male prudently flees. Returning to Zhen, the large male mounts her, she crouches and he squats upright by her rump, his forepaws resting on her lower back. However, she slips out from under him. The small male comes near, moaning, and is promptly attacked again, though I only hear growls, roars, and whines like a pack of dogs fighting and see the bamboo shake violently. Back in the open, the big male mounts Zhen again, and the small one, still undeterred, twice briefly approaches the pair. Apparently deciding that the chances of fulfilling his sexual imperative are slight, though not quite zero, the small male drifts aside and waits. So intent have I been on observing and tape-recording, that I have been unaware of Zhu Jing's and Zhou Shoude's arrival. There is now a lull in activity, and they continue uphill to monitor for the night.

Zhen suddenly climbs a hemlock about fifty feet from me where she casually sits on a low branch, one leg dangling free, looking at me blankly. The male soon comes and circles the tree bleating; she peers down at him and hoots, safe from his determined advances. With a swaggering sense of invulnerability the male ambles to within twenty five feet of me and there flops on a bed of moss, panting heavily, lying so relaxed he looks like a huge animated cushion. I debate what to do. A male's social interludes are few and seldom amicable, and there is a suppressed air of ferocity about him as he guards Zhen. Will he treat me as another male panda? He rises and approaches, his round white head having an almost hypnotic effect as it pushes closer and closer through the bamboo. Standing motionless, I wait. He halts eight feet from me and merely looks, his face a mask. Returning to Zhen, he rears up against the tree as if to climb it while she leans down and hoots. Xiao Wang, returning from the ridge, joins me in time to see the female descend.

I have had various unusual meetings with animals, each in its own way deeply satisfying. A curious female mountain gorilla climbed a tree to sit on the same branch with me; a snow leopard remained by her kill during a windswept night in the Hindu Kush even though I rested nearby; and a tigress and I met inadvertently in the jungles of central India, my eyes meeting her golden gaze at little more than arm's length. And now, for the next hour, Zhen and her consort provide a unique glimpse into their inscrutable lives. Zhen comes to the other side of the fir by which we stand and stops six feet from us, only her head visible, gleaming in the forest like a full moon on a frosty night. Xiao Wang steps behind me. I flick one hand at Zhen and say softly "tch-tch" to alert her to our presence, but her trance-like expression does not change. She acts as if Xiao Wang and I are invisible, transparent. She slumps against our tree like a sack of flour. Afraid of being in the male's path, we sneak around the other side of the tree and sit down fifteen feet from Zhen. Film in my camera finished, I need another roll from my rucksack, which is beside Zhen. I crawl to her, cautiously, and retrieve it without eliciting a response. Each creature has its own window on reality, and my encounters with pandas have been too brief and rare to gain insight into what took several million years to evolve. But I have never before met a pair of mammals who filtered out past, present, and future so completely that only one focus, one reality, remains.

When the male steps from the bamboo, Zhen approaches and crouches with her muzzle tucked so far under that the top of her head rests on the ground. He mounts, dismounts, and slowly circles her immobile form, and mounts again, and twice more. They then walk a little, Zhen leading, both bleating, he wholly engrossed as he mounts, usually for ten to sixty seconds, whenever she stops; at times he paws her softly or maintains body contact by leaning against or over her. Once, when he tries to stop her restless rambling by grabbing one of her hindlegs with his forepaws, she whirls and emits harsh barks. Another time when he importunes too persistently she simply climbs a spruce to escape his ardor. Once the male walks carefully around my tape recorder on the ground as if it were a stone, and mates with Zhen only ten feet from us as if we are of no more consequence than a couple of logs.

It is nearly dark at 7:15 P.M. During the past seventy minutes the male has mounted forty-two times, and the female has become more quarrelsome and less receptive to his advances as they move slowly back into the depths of the bamboo. And there the second male continues his fruitless vigil.

14 April. Zhen lazes around the swale where she has courted. Kay and I wait nearby, hoping to glimpse either her or one of the males. It is a radiant spring day of a kind rare at Wuyipeng. Rhododendron leaves glisten and buds are swollen and pink, ready to burst into flower. The red bark of birch catches the sun's slanting rays until the forest is filled with columns of fire, and lichens droop in luminous strands from boughs. Zhen sleeps the afternoon away. Of the males there is no sign. They obviously seek sexual heat rather than social warmth. Zhen's estrus peak has lasted less than a day.

Her courtship has some features that captive pandas do not or cannot express. The male calls loudly, sometimes from up in a tree, a love song of sorts that advertises his presence and readiness to mate. Any female can then seek him out, if she is so inclined. Zhen remained near the male but did not join him until she was ready, behavior that ensures more-or-less amicable relations, quite different from the abrupt and forceful way in which captive pandas are often introduced to each other.

It had been a disjointed few weeks, the days at Wuyipeng interspersed with two visits to Yingxionggou and a week of project meetings in Chengdu, and on 16 April we received a message that Emil and Stephen would arrive from Chengdu for another brief stay. Kay and I descended to headquarters to meet them. Village women were carrying baskets of manure into the fields, and men moved up and down the steep hillsides behind wooden plows pulled by yak-cattle hybrids; apple and almond trees were in bloom around the houses. In our room at headquarters, I placed bamboo samples on a low table to dry above a hot plate. While we were bathing, a power surge ignited the bamboo, which then set fire to the mattress on the bed. Returning to the room, I found the mattress a wall of flame, and in the fire was the camera bag I had left on the bed. A bucket of water extinguished the blaze. Lost were the bamboo samples, a mattress, and a camera with two lenses—but not the photos of courting pandas or a set of field notes which I had taken from the bag minutes before. (As a safety precaution, I always keep two sets of field notes, each in a different location.)

The following day we had a meeting to discuss work, and then a day at Yingxionggou where Emil and Stephen electro-ejaculated two males and froze the semen in liquid nitrogen for future use. On our return to headquarters, I was informed that a panda had been caught in a trap at Wuyipeng—a total surprise, for I was unaware that any trap had been reopened and baited.

We reached the captured panda toward noon the next day, high on the ridge where Zhen was first caught. It was a young female similar in age and size to Long. Extraordinarily tame, she took bamboo from our hands and ate it. When Emil scratched her head she leaned toward him, and she extended her forepaw for us to hold and stroke. But neither she nor other pandas liked their hindpaws touched. Emil sedated her. Within an hour she had trotted back into the bamboo wearing a radio collar. We named her Ning-Ning, meaning "gentle" and "kind."

Toward the end of April, the forest became resurgent with life; spring was at its peak. Thickets of rhododendron laden with lavender blossoms surrounded our tent, and glowing in the interior of the forest were the pale pink flowers of a tree rhododendron. The *Tetracentron* tree, massive and gnarled and often hollow, the perfect retreat of gnomes, now had amber-colored leaves, small heart shapes that caught the first sun from the eastern ridge to shine like a constellation of golden stars strewn against the morning sky. The *Tetracentron* was once part of an ancient ecosystem in the western Chinese mountains that remained secure and isolated one hundred million years ago while the region that is now Tibet was covered by the Tethys Sea. Then when vertical uplifts and horizontal thrusts pushed the floor of the Tethys Sea skyward to create the Tibetan highlands and later the Himalayan mountain system, new migration routes opened to the ancient flora. Some plants—the rhododendrons, oaks, and bamboos—emigrated far into other areas, but the *Tetracentron* remained, a golden relic at Wuyipeng.

Maple and larch burst into leaf, imparting a soft, green sheen to the lower slopes, and fern fronds pushed up through the mat of sodden leaves. Several migrants returned, among them Gould's sunbird, brilliant in scarlet and yellow. Golden monkey groups that in winter were widely scattered, searching for their frugal diet of lichens and birch buds, now joined and together foraged on nutritious young leaves. One day over two hundred monkeys passed around me, males and juveniles and females with infants, the air vibrating with their cries and the trees seething and rocking as in a storm. With the warmth a few land leeches came out too. Although they liked me well enough, I somehow found them inaccessible to my sympathy. When three people walk in the forest, Hu Jinchu confided, it is best to seek the middle, for the person in front gets bitten by ticks and the person in back by leeches. On 20 April, I noted the first new shoot of umbrella bamboo, almost an inch thick and covered with rust-colored whiskers.

Zhen moved down from the ridge at the end of April, away from arrow bamboo into umbrella bamboo to feed on the juicy shoots. Bamboo reproduces annually not by seeds but by sending up shoots from underground stems, or rhizomes. Zhen and several other pandas had adjusted their schedule so as to be low on the slopes during this seasonal feast. In winter the pandas ate mostly leaves and young stems of arrow bamboo. But starting in April, they suddenly disdained leaves and concentrated on crunching dry, tough stems, a diet they would favor until late June when they switched wholly to leaves, and then ate almost no stems until November. Since old stems are woody and low in nutrients, I wondered why leaves are unpalatable for those three months. Chemical analyses did not provide an answer and the puzzle remains unsolved. However, I could at least appreciate Zhen's preference for adding shoots at this season.

Because we were no longer trying to trap pandas, there had been an exodus from Wuyipeng. Coworkers returned to their home institutions after brief tours here, attended meetings, went on leave, or otherwise departed, leaving only Xiao Wang, Kay, and myself. Kay and I concentrated on studying shoots while Xiao Wang radio-tracked Long, Zhen, and Ning. To determine how fast shoots grow, we measured certain ones every evening. At first they averaged about an inch of growth per day, but in mid-May, after the onset of warm, wet weather, they shot up at a rate of three inches a day and some as much as seven inches, to reach their full height after forty to fifty days. We often came upon a panda's feeding site, a litter of whitish sheaths that had been stripped off the shoots and discarded. By measuring the diameter of shoots eaten and not eaten, we found that pandas prefer thick shoots, those more than a third of an inch in diameter, no doubt because it took a panda as much effort and time to remove the sheath from a thin shoot as from a thick one but with less return in amount of food. Pandas often foraged along the edge of bamboo or in small clumps rather than deep inside thickets. Why? Shoots grew more densely near the edge and there were proportionately more thick shoots there. We counted the number of shoots a panda had eaten and not eaten in several stands and found that in one visit a panda might take fifty to seventy-five percent of the shoots available. Since each shoot eaten means one stem less, pandas obviously had great impact on bamboo. The more we looked at shoots, the more questions we had. We saw that shoots were so abundant that a panda could plunk itself in one spot and eat at least three to four without shifting. But how many shoots, how many pounds of shoots, did Zhen eat in a day?

With direct observation of pandas in the wild difficult to make, I went to Yingxionggou to learn more about shoot-eating from captives. One male there was called Ping-Ping, meaning "peace," a most inappropriate name though not through a defect in his character. Dogs had ripped off his ears during capture. He had adapted poorly to the diet of gruel, which he did not like and which his digestive system barely tolerated. Since there was never enough bamboo, he suffered from chronic malnutrition. These problems made him irritable, and because keepers teased him he had become aggressive. I tried without success to have him released in a remote part of Wolong. Aside from making Ping-Ping happy, we could have learned how quickly and well a panda would adapt to new terrain. One reply to my suggestion was, "He would make a good anatomical specimen." Later Ping-Ping badly injured a careless photographer at Yingxionggou. Pandas project such charm and vulnerability that people tend to see only the image, not the animal—a panda is powerful, potentially dangerous, and its moods are understated. Keepers have been seriously mauled in the Chicago, London, and Washington zoos. Since only thirty-eight pandas lived in zoos outside China between 1937 and the early 1980s, pandas have inflicted proportionately more serious injuries than any other captive species except elephants.

All through the day I fed only shoots to Ping-Ping. Sitting close to the bars where he could take each shoot as I handed it to him, Ping-Ping crunched down shoot after shoot with happy abandon. Discarded sheaths soon covered his lap and legs. When satiated, he rolled on his side and napped, so abruptly it seemed as if a switch in his brain had suddenly been turned from "eat" to "sleep." While he rested, I collected more shoots on the nearby slopes and weighed them, and raked in his droppings and weighed those. Soon Ping-Ping was ready for more, his stomach empty. Pandas can process shoots in as little as five hours between eating and defecating. Ping-Ping required an average of only thirty-seven seconds to peel and eat a shoot, and in the course of a day he consumed thirty-six pounds. When I left him after two days, as he expectantly waited for me to bring more of this treat, I was saddened that his interlude of pleasure had been so brief.

If Ping-Ping's thirty-six pounds of shoots a day indicated a healthy appetite, then Zhen's was gargantuan. Once Zhen spent about ten hours in a bamboo patch and in that time ate two hundred and eighty one shoots—about one every two minutes including rest periods—and deposited fifty-seven droppings. Knowing from Ping-Ping the conversion weight of shoots to droppings, I could calculate that Zhen had eaten a total of about seventy-six pounds of bamboo in that twenty-four-hour period. On other days she ate more, for an average of eighty-four pounds (thirty-eight kilograms) a day, or nearly half her body weight. Because a bamboo shoot consists of ninety percent water, she needed all that bulk just to meet her daily nutritional requirements. She was somewhat like a person who subsists only on watermelon. According to Ellen Dierenfeld, the New York Zoological Society's nutritionist, bamboo shoots possibly serve another important function as well. They contain a compound, pithily known as 6-methoxybenzoxazolanone, which is known to trigger reproduction in rodents. Does it have such a stimulating effect in pandas too?

Did a panda spend less, more, or the same amount of time each day foraging for shoots as for stems and leaves? With Zhen near camp, we could easily monitor her from our tent. Kay usually had the first night shift while I slept. Then I took over until morning, sitting quietly at my desk. A fire burned in the stove, for the May nights were still cool. Rain pattered on the roof. I was told the rainy season would not start until the end of June, but already dark clouds swirled around our valley and everywhere water splashed, trickled, and dripped. The kerosene lantern cast a soft light. Hidden in the darkness, a tawny owl hooted. On the desk before me was the receiver tuned to Zhen's radio frequency, the disembodied signal telling me that she was intermittently active at night, just as she had been earlier in the season. The hours were long, filled with reading and writing and drinking tea. But they were not lonely, because Kay was there only a breath away, and *Apodemus* mice scurried boldly around my feet in such numbers that word must have gone out that our tent was the residence of choice. At dawn I stepped out to look at the weather. The air was quiet in the fog, and the lavender blossoms of rhododendron had a cold sheen as if made of ice.

While Kay continued daytime monitoring, I went into the forest. The boughs and bamboo dripped as I knelt on the ground to look closely at a panda feeding site among ferns, snakeroot, cowslip, nettle, false Solomon's seal, and other plants I knew from New England's woods. I examined and measured the panda's leavings, collecting any scraps of information that might help answer the many questions that crowded my mind. Perhaps to most persons my fascination with bamboo would seem to be of as little value as the art of butchering dragons, and as exciting as watching shoots grow. Although facts in themselves have little meaning, they are points of departure for reasoning and ideas, for seeing things in a new way, in this case for understanding the panda and its natural world.

I often wished for a Chinese coworker with whom I could discuss research methods and share knowledge and ideals. That was part of the reason for coming to China. But the five permanent research workers in camp seldom did anything beyond the routine, such as radio locating the pandas, and they asked few questions. Camp was not imbued with the Confucian dictum that "Extension of knowledge lies in the investigation of things." An interesting cyclical phenomenon occurred once a month when upset stomachs, lame legs, headaches, toothaches, and other afflictions incapacitated workers just before twenty-four-hour monitoring of the pandas' radio signals was to begin. I sympathized with the workers' preference for drinking tea and listening to daily soap operas on the radio; after all, workers were paid only a small bonus for living under these miserable conditions, and they derived no benefit from the information they collected. I knew that one can demand work from but never do good science with disinterested people. Or, as I once told Wang Menghu: "There is an aphorism in America, 'You can lead a horse to water but you can't make it drink'." I needed creative and resourceful coworkers, people of a kind that the Chinese system tends to suppress. A scientific symposium on the giant panda, published in 1974, opened with an editorial entitled "Deepen criticism of Lin Piao and Confucius and persist in socialist revolution," an exhortation that probably had little influence on improving standards of scientific research. At any rate, I found it frustrating to be essentially a guest, unable to order or delegate work. To give anything resembling an order was considered arrogant—a serious accusation—as was pointing to an error in someone else's method of collecting data and imparting knowledge without at the same time being self-effacing. Already I was considered "proud" and "bossy" because, for instance, I had sent Howard Quigley a telegram saying he should not radio collar giant otters in Brazil, a message which at Wolong headquarters was somehow translated as meaning that I forbade Howard to marry! While such an incident was superficially trivial and humorous, it did shape perceptions.

One day, when Hu Jinchu was on leave, I outlined various problems, some challenging, some irritating, to Zhu Jing of the Academy of Sciences, who had come on a brief visit. I noted that our equipment was being ill used, cameras and telemetry items left in rain and dirt. It was necessary to appoint someone to care for equipment, because if no one had responsibility everyone shirked it. I indicated that most research was being done by Hu Jinchu and myself and that others should share more of the tasks. Furthermore, work was often carelessly done, sometimes so poorly that data had to be discarded. I noted these things and others, and Zhu Jing, current in science and fluent in English, listened carefully; just as carefully he dissembled, indicating that he could not respond, and he did it very well. Later I told the same to Hu Jinchu. He replied gently:

"I have also seen many problems in the spirit of work. But I can do nothing. Like you, I am only a guest in Wolong. I have the title of leader but not the authority to give orders. I cannot prevent others from leaving or make them work; I can only make suggestions and the others will decide whether they wish to follow or not. This is our system, but it is changing."

I slowly came to realize that Hu Jinchu's position was far more difficult and frustrating than mine. Furthermore, he was hampered by the philosophic Chinese attitude of preserving appearances. I continually asked him questions about natural history, my culture being tolerant of a display of ignorance, and he generously shared his knowledge gathered during years of tramping through Sichuan's mountains. But he could only be restrained about seeking information from me. Being a professor, full of pride and dignity, he was expected to impart knowledge, not reveal a lack of it, especially not in front of the interpreter whom we needed for such discussions. That, at least, was my interpretation of the situation, not necessarily the correct one. Austin Coates expressed my feeling of confusion in his book *Myself a Mandarin:*

> To anyone trying to understand other races in this world, the Chinese surely pose the greatest challenge; and they too have their own difficulties in trying to understand the rest of us. For my own part, I found the pursuit of understanding them was like a game in which I was always the loser.

In every project there are periods of soaring spirit or of depression. In mid-May, unable to immunize myself against problems that had multiplied since November, like the mice in our tent, I succumbed to gloom. I had fumed impotently, caught in the intrigues and tensions of a society shaped by the Cultural Revolution which although it was officially dead was not yet buried. It was a shadow world of manipulated facts and evasions. An open scientific approach to the project did not exist. Relevant Chinese publications on pandas were deliberately withheld from me, as were laboratory analyses of bamboo that had been done from samples we collected. My requests to take bamboo overseas for a kind of analysis that could not be done in China were viewed with the utmost reluctance. Were data being held hostage until WWF provided all equipment for the research center? Did the Chinese think I would abscond with data? I was irritated more by the silliness of the acts than angered by the lack of trust. But as in many things, it is the thought rather than the act that hurts, and I noticed in myself a gradual death of pleasure in the work. I felt myself disoriented and lacking, inadequate to handle this project. In my previous research I knew where I was going and why; I had control. Not here. Restraint can be menacing. I had worked with a great sense of urgency. After all, with the panda it would not be enough to *try* to save it: there must be success. And might not the project be suddenly stopped, as the Chinese had threatened this past November?

There was also the pressure to do well, for every scientist competes not only with himself or herself but also with all previous standards of excellence. My remorseless curiosity and single-minded purpose had created mistrust in a culture where one triumphs over existence by accepting it, by flowing with the stream. I felt my validity diminished. The weather did not help my mood. A dark tapestry of cloud obscured the hills. Camp was a squalid bog. The Chinese workers felt dejected too. Constant shifts in personnel had disrupted congeniality, and everyone now ate hurriedly and alone. In a few days Kay would leave for America to spend the summer with our two sons, who were returning from university, taking with her the kind words and caring smiles that provided such solace in this place so stifling with silent tensions.

Lying in bed at night, Kay and I talked with nostalgia of past projects. "Do you remember . . ." one of us would say. And then we relived our climbs in search of mountain gorillas among volcanoes, foggy and mossy and steep, with bamboo low on the slopes much like those at Wolong.

"It's so beautiful here," said Kay. "Wouldn't it be wonderful if we could do this project alone? Just the two of us. Just as we once did. With only ourselves to please. It could be so perfect."

And we talked of our years in the Serengeti, of driving into the golden dawn and finding a group of tawny lions beneath an umbrella acacia. And we talked of tigers in India. And of Brazil and Pakistan. "Do you remember . . ." We had a habit of thinking in terms of freedom, freedom to plan

Zhen-Zhen strolls among umbrella bamboo near our camp.

The Wuyipeng research camp huddles on a mountain slope in a forest of bamboo, rhododendron, and birch.

October snow on maple leaves heralds the onset of winter.

(facing page) Tang-Tang awakens from his rest in a pine in the Tangjiahe Reserve, startled to observe me nearby.

(facing page) Vertical mountains, clouds, gnarled trees—the Wolong Reserve is the landscape of a Chinese brush painting.

(right) Protecting the panda also gives thousands of other species such as these blood pheasants a future.

A Christmas banquet at Wuyipeng. From left, facing the camera, are Kay Schaller, Bi Fengzhou, and Qiu Mingjiang

Howard Quigley measures the sedated Zhen-Zhen while I take notes.

Teng Qitao (center), leader of our Tangjiahe team, helps raise the door of a transport cage to release Tang-Tang back into his home after radio collaring.

A subadult male, wearing his new ear tag No. 86, recovers from sedation inside the log trap in which we caught him.

Teng Qitao (with antenna) and Shen Heming monitor radio-collared pandas and bears in the Tangjiahe reserve.

In June, *Pleione* orchids bloom on mossy boulders at Tangjiahe.

our daily lives, to come and go and to express opinions openly, to dream and transform our dreams into reality. We were spoiled, and this sometimes made it difficult for us to adapt continually to a society in which Confucian tradition stresses respect for hierarchy and the common good rather than prizing individual desires and a free spirit. And, in addition, it was useful for us to remember that China had traditionally tried to segregate and control all foreigners within its borders. When, for example, Emperor Kang Xi initiated his open-door policy in 1685 by permitting foreign trade at all Chinese ports, soon there were so many rules and regulations, the movements of visitors so restricted, and demands for presents so exorbitant that commerce was quickly stifled.

Fortunately we usually took turns at being depressed so that the other was there to add perspective.

"It's the lack of freedom that's so difficult. If we could only take advantage of a slow spell to get a break in routine and go somewhere else for a few days. But by the time we could get permits it would be too late. I'd be happy if we just had a day's picnic on the spur of the moment."

"I know. But even if the mistrust and indifference are hard to take, remember that the Chinese treat each other no differently. And we're here by choice; we really like living this way."

"And we have an option they don't have. We can leave."

So, with a sense of humor and an appreciation of surrealism, I would overcome my weariness. I would seek the freedom of camp life yet feel caged, work hard in a place where idleness was a virtue, be invited to do research yet be silently discouraged from succeeding, work on behalf of saving a beloved world treasure that is locally treated with indifference.

I would submerge my individuality and accept restrictions—but I would not become resigned.

High up on the ridge, birches were just coming into leaf in late May and early June, and the slender shoots of arrow bamboo pierced up through the moss. But around Wuyipeng spring was over. On my daily hikes in search of pandas, the forest around me was lushly green under a rain sky. Petals of wild cherry drifted down like belated snowflakes. Flowers were inconspicuous, those of trillium, *Androsace*, pyrola, and others a subdued white in the rank growth. Tragopan chicks had hatched, the mother pheasant clucking as she led tiny puffs of down through thickets. At trail side was an occasional viper, *Trimeresurus*, handsomely speckled in yellow, brown, and greens; and skinks—small dragon-children as the Chinese call them—scurried away.

27 May. One day I follow the contours of a slope, my footsteps silent on the spongy ground. My receiver is tuned to 194, Zhen's frequency. The signal is so loudly insistent that I expect to see her standing by a nearby clump of bamboo. My eyes try to penetrate the stems, but I see only shades of dark, moss-covered rocks and tree trunks and the dusk in between. I sit down in a small clearing and wait. Immobile clouds hang in the valley, hemlock boughs sag with rain, and even the still air is melancholy. Minutes pass. I try to imagine a panda's life in the depths of bamboo, always surrounded by stems, enclosed in a dome of leaves pressing close, and beyond the white of fog. Every touch of a bamboo stem releases a shower; if pandas had a theme song it would surely be, "Raindrops keep falling on my head."

Sounds downhill suddenly transform the forest—the snap of a breaking bamboo shoot, followed by the rustle and squeak of sheaths being torn away, and finally loud crunching as Zhen eats the tender center. Minutes pass to the sounds of her foraging. A panda's sense of smell is acute. Would she notice my presence and vanish unseen? But no, she angles noiselessly uphill. I detect a fleeting movement, then watch raptly as she emerges and sits behind a thin screen of bamboo. She leans sideways to hook in a bamboo shoot with the curved claws of a forepaw, and with the same motion breaks off the shoot near its base. Then, settling back and holding the shoot at a slant, she bites into the sheath and with mouth pulling sideways and forepaw jerking down while twisting, she tears away the sheath and drops it. She then takes a few quick bites, pushing the shoot into the corner of her mouth where it becomes rapidly shorter like a pencil in a voracious sharpener. Glancing around, she spies another shoot; peeling and chomping, she eats it too within a minute. A third shoot follows, her actions calm and deliberate in complete accord with her surroundings, yet at the same time fluid and rapid, as if she has little time.

As I watch her eat, I am impressed by her dexterity, forepaws and mouth working together with great precision and economy of motion. Evolution has provided the panda with special adaptations for subsisting on bamboo. There is a sixth finger, a prehensile elongated wrist bone, the radial sesamoid, that functions as a thumb, ideal for handling shoots and stems, even those of arrow bamboo a fraction of an inch thick. Stems are held as with forceps in a hairless groove connecting the pad of the first finger and the "pseudothumb." The panda's typical carnivorous dentition has been modified for crushing and grinding tough food; not only the molars but even some premolars are broad and flat. The skull is unusually wide and there is a bone crest on top of the braincase that provides a point of attachment for the powerful jaw muscles. The panda is a triumph of evolution. However, by becoming highly specialized on bamboo, it has reduced its options. Freed from choices, it may seem at first glance to be freer than most animals, yet evolution has robbed it of innovative vigor, imprisoned it with an ecological possessiveness that knows no reprieve: bamboo has fettered the species with its impassive power. Although I marvel at Zhen's evolutionary specializations, I am also touched by the tragic history of her species, her helplessness. Indifferent fates have mastered her.

Zhen raises her nose as if testing the air, apparently sensing my presence. With a rolling motion she rises and moves from behind the bamboo into a bower from which a path leads to my clearing. She steps forward with a combination of shyness and audacity. Her black legs dissolve into the shadows to create an illusion of a shining lantern gliding toward me. She advances to within thirty-five feet. There she stops, her head bobbing up and down as she snorts to herself in wary alertness, her apprehension and mine a bond of shared feeling. I look for some intimation of coming actions in her face, but it remains inert, showing neither passion nor docility. Here the panda does not invite familiarity but is imposing with the same durability as the fir trees and rugged peaks, complete in itself, final.

Not being a creature of self-expression, Zhen conveys none of her inner feelings. I wonder what she will do now. Certainly she does something unexpected. After gazing intently in my direction, she retreats to the margin of the bamboo and there, slumped against the stems, gives agitated, bleating honks, a strangely vulnerable and timid sound from so large an animal. Hunched in repose, forepaws on rounded abdomen as if meditating, she has the aura of Buddha. Her honks grow softer as her head sags onto her chest. The slow rhythmic heaving of her body reveals that imperturbably she has fallen asleep.

Although Zhen has just eaten and is therefore programmed for a digestive trance, I hardly expect her to fall asleep in front of me. What intuition, what reason is there in that broad, hard skull? A panda has its version of the world and I have mine. What does the world look like to a panda? On meeting a gorilla or a tiger, I can sense the relationship that binds us by the emotions they express, for curiosity, friendliness, annoyance, apprehension, anger, fear are all revealed by face and body. In contrast, Zhen and I are together yet hopelessly separated by an immense space. Her feelings remain impenetrable, her behavior inscrutable. Intellectual insights enrich emotional experiences. But with Zhen I am in danger of coming away empty-handed from a mountain of treasure. To comprehend her, I would need to transform myself into a panda, unconscious of myself, concentrating on her actions and spirit for many years, until finally I might gain fresh insights. Yet I see Zhen and her kind so seldom that, although I can capture a little of the science of the animal, her being eludes me. I am not even certain how to begin. The panda is the answer. But what is the question?

Zhen soon awakens, and without hesitation or a glance in my direction she moves uphill, drawing the shadows around her so abruptly that her dissolution is as startling as her appearance. I remain sitting, unwilling to disturb her, wherever she is, and wishing to prolong these moments. Rain now whispers on the leaves and there is a sound as of distant surf in the tree tops. How long have I been in the clearing? Time has no single measure. Absorbed in Zhen, I am released from past and future until she breaks the spell of our meeting, leaving me with feelings stronger than any memory.

By mid-June most bamboo shoots had grown into tall, hard stems, and soon Zhen moved back up into arrow bamboo. There she would remain until next year's shoot season once again enticed her down.

I was suddenly given permission to export bamboo samples for analysis. The inappropriate gratitude I felt on receiving this news surprised me. However, I assumed that the granting of this request also signaled that I would be allowed to reenter China. Needing to rekindle my enthusiasm and joy in the work, and wishing to see my sons as well as attend to office work, I planned to go home for two months. Zhen had mated in mid-April, and her baby was due sometime in September. I wanted to be back for that event.

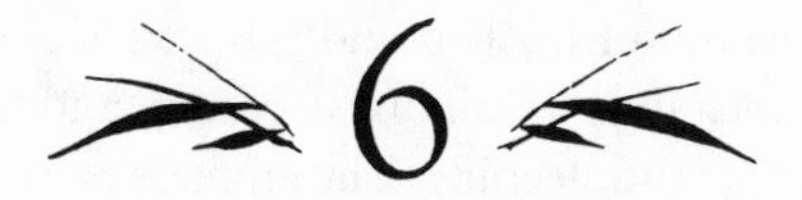

In The Hollow of a Fir

September to December 1981

When I returned in early September, maize cobs had tassels and beans were ready for harvest in the fields of Wolong. But at Wuyipeng nothing seemed to have changed. A cover of immobile cloud still imparted an aura of gloom, trees dripped, trails were muddy and slick as ice, and rivulets like silver scrolls cascaded down slopes. Yet subtle changes marked the advance of the season since June. Only a few flowers, a yellow *Pilea,* a blue aster, were at trailside, but many kinds of mushrooms—purple, scarlet, pink, white, sulfur—illuminated the somber forest. Tragopan chicks had become plump pheasants almost as large as their mothers. Rock squirrels stuffed themselves with hazelnuts in preparation for a winter of hibernation. Three of these squirrels, chocolate-brown, pointed-nosed, lived at camp. When we held out a peanut, a squirrel would cautiously approach, tail nervously flicking, dart up our pantleg, grab the food, and scurry to safety. These and other signs revealed that autumn was near.

On arrival at camp, I inquired immediately about Zhen. In mid-August, I was told, she had moved across the ridge from the Zhuanjinggou into a dense expanse of bamboo just above Erdaoping where there were two suitable den sites, both in fir trees. A female needs a protected den for her young, but suitable sites are scarce and there may be none in logged areas. A fir, for example, must be at least three feet in diameter—such a size requires two hundred years of growth—and have a hollow base with an entrance large enough to accommodate an adult panda. No one had checked on Zhen for fear of disturbing her, and it was not known if she had given birth, or for that matter if she was indeed pregnant.

Hoping that Zhen's daily cycle of activity, as shown by the radio signal, would provide clues about the status of her motherhood, we began monitoring her day and night, listening to the signal at fifteen-minute intervals from 4 September until 24 September, more an exercise in endurance than pleasure. In previous months, Zhen and the other pandas were active for about fourteen hours a day, usually busy eating, and their inactive hours were devoted to resting and sleeping. The animals had one or two long rest periods lasting two to four hours, occasionally as long as six hours, each day. They might be active at any time but in general had a daily cycle: they were most likely to move around just before and at dawn from 4:00 to 6:00 and in late afternoon from 4:00 to 7:00, and they were most inclined to rest soon after daylight between 8:00 and 9:00 and around dusk after 7:00. They usually ignored bad weather, except that an animal might shelter beneath or inside a tree during a heavy downpour. Pandas rested a bit more in summer, when they could quickly fill themselves with leaves, than during other seasons when they ate many stems, which first had to be selected and then laboriously chewed down piece by piece. We scanned Zhen's activity schedule with great interest at the end of each day's monitoring. It differed greatly from her previous routine. She was far less active, moving around for only about nine hours a day, she lacked daily activity peaks near dawn and dusk, and she was extremely restless, as shown by the many changes in signal speed. Furthermore, she remained within six hundred feet of a den tree, apparently making only brief trips away from it to eat and drink. We were now certain Zhen had a baby, probably born on 2 or 3 September, a day or two after she had settled in the den area. The gestation period had lasted about one hundred and forty-two days.

Gestation periods in captive pandas vary from ninety-seven to one hundred and sixty-five days, with an average of about one hundred and thirty-five days. This variable period indicates that pandas have delayed implantation, in which the fertilized egg divides only a few times to the blastocyst stage and then floats free in the uterus for some time before implanting and developing further. This system is found in various carnivores, including bears. In pandas the delay in implantation varies from about one and a half to four months, and the young then actually develops for only one and a half to two months, when it is born in a physically undeveloped state—blind, nearly naked, and almost helpless. It weighs only three to four ounces, no more than a rat or chipmunk, one nine-hundredth the weight of its mother. If the same size relationship existed in humans, a one-hundred-and-thirty-pound women would give birth to a baby weighing two to two and a half ounces. I would have liked to observe the behavior of bulky Zhen with her minute infant, and I think she would have permitted my presence, quiet and distant. However, with a dozen others in camp who would also want to visit her, the disturbance might cause inadvertent harm to the infant. When Mexico's first baby panda was born, so many photographers crowded around that the female became nervous and accidentally crushed her newborn. One day when Zhen's presumed young was around three weeks old, and would weigh about two pounds and look like a miniature adult, judging by the development of zoo animals, Hu Jinchu suggested that he and I verify the presence of a young panda to make certain that our radio monitoring was not a futile exercise. I agreed. However, Long Zhi, a temporary researcher from the Academy of Sciences said: "The female might eat her young if disturbed." But the decision whether to go was mine, he noted, and mine alone. I declined this responsibility, knowing that I would be blamed if at some stage in its life the young disappeared. We would wait.

My daily routine, like Zhen's, was now mundane, my journal of events uninspired and laconic.

14 Sept. Rain much of night. Spent morning fixing things up around tent—dug drainage ditch, made stone steps on slippery incline. My turn on the night watch. Left at 1415. Saw the golden monkeys on the way. Chilly; in the 40s. Fog and rain.

15 Sept. Off at 0730 to do azimuths. A strong wind from up-valley tears fog apart, until lower hills are clear. But tops of mountains still solidly in cloud. Vegetation is getting a tattered look—leaves full of insect holes, some brown-edged, dying. Home at 1000. Slept 3 hours. . . . Walked to Bai Ai at 1600 to look for monkeys. No luck.

16 Sept. Weather continues gloomy and wet. I work at summarizing notes to save time next year. At 1600, I take a walk past Bai Ai. Seldom see wildlife except an occasional tragopan female with young.

17 Sept. Shortly after 1000 went to Monkey Rock area and there circled through the bamboo, looking for fresh panda sign. Three den trees in the area and I check these. Nothing.

18 Sept. A clear morning! It is my turn to do the daytime monitoring on the ridge. The forest is shining, and, instead of crawling into the sleeping bag to stay warm, I stand in patches of sun. Flies buzz. It got steadily cloudier and by 1600 it was overcast, but just to see the mountain peaks gave a lift to morale. . . . News from Chengdu: a baby in zoo. I must go and observe it and see Janet Stover.

Janet Stover, a veterinarian from the New York Zoological Society, had arrived in late August to help hand-rear newborn pandas at Yingxionggou and Chengdu, if a female failed to care for her young. If twins are born, a female usually abandons one, and to hand-rear a tiny newborn is a delicate task that requires devotion and knowledge. For example, a young is unable to urinate and defecate unless its belly is first stimulated by licking—or by something equivalent—and it needs a milk formula with the correct balance of fats and sugars. Unfortunately Janet was unable to apply her skills because no young were born at Yingxionggou, and Chengdu's Mei-Mei had only one young.

Mei-Mei had given birth at 2:00 in the morning on 18 September to a male named Jin-Jin, and I joined Janet for three days of observation on 22 September. Sitting on a bed of hay, Mei-Mei held Jin-Jin with one paw to her chest. So broad and hairy was her paw that the baby was completely hidden from view, and indeed I would not have known of its presence if it had not given a harsh, penetrating squawk. Grasping Jin-Jin lightly with her jaws, mouth over his back, Mei-Mei picked him up and leaned forward with the suppleness of a gymnast to rest her head on her hindleg, using it as a pillow. Briefly I saw Jin-Jin, a pink body covered sparsely with a silver sheen of hair, a bare rat tail, and spindly legs, resembling an ill-designed rubber toy, before he disappeared from view, tucked beneath his mother's chin and covered by a paw.

Once when Mei-Mei shifted, he emitted a sharp squawk and she adjusted her arm. When your mother is so much heavier than yourself, it pays to have an outsized call to advertise your presence. Jin-Jin used it emphatically whenever he was uncomfortable, which was quite often, as when his mother held him upside down, let his legs dangle, or picked him up carelessly hard with her teeth. About once an hour Mei-Mei solicitously licked his chest, back, belly, and rump for several minutes. We were told that Jin-Jin suckled approximately ten times a day on one or more of his mother's four teats, but he was so well concealed in his mother's paw that we could not observe what he was doing from outside the cage. Mei-Mei frequently shifted position just as Zhen did, sitting, lying, turning over, yet she never put Jin-Jin down; even when she left her bed to relieve herself she carried him along in her mouth. Not once during the first twenty-six days of his life was Jin-Jin permitted to leave her arms. By then he was well furred, less in need of his mother's warmth and protection, though he was still helpless. His eyes would not open fully until the age of one and a half months, and he would not be able to take shaky steps until two and a half months. Seeing Mei-Mei's almost marsupial devotion, her paws serving as a pouch for her offspring, I could understand why a panda cannot raise twins easily. No female in captivity reared twins until 1990, when one female in the Chengdu Zoo did so. And that same year, twins about three weeks old were discovered in a cave in Wolong.

Like all field biologists, I often lead a lonely life in the wilderness for love of the work and the touch of adventure that any project can provide. There is joy in hardship, in overcoming obstacles, in pitting oneself against the elements. However it is one thing to embrace discomfort when the sun is shining and quite another when the weather is bad. As September turned into October the valley was still filled with dispiriting clouds that enshrouded us in twilight, damp and chill. There can be radiance in melancholy, as when at autumn's end maple leaves turn to crimson and fall. But I found no such radiance in this gray fog. Only minor interludes interrupted the tedium of daily routine. On 4 October a large pile of mail was delivered, and I read letters and magazines far into the night. At least with today's ease of travel, a letter to America could bring a response within six weeks; by contrast, when the Italian Jesuit missionary Matteo Ricci sent a letter home from China around the year 1600, he would not expect an answer for six years. The gray days continued and on 8 October I awoke to tent and forest sagging beneath three inches of sodden snow.

Finally, on 15 October, the clouds were high and there were patches of blue. I was seized with a desire to go, to go somewhere, anywhere; I yearned for the anarchy of being free. After a hurried breakfast, I set out for the eastern rim of the Choushuigou, an area I had not previously visited. I climbed up a long spur to the point where bamboo and fir gave way to a slope deep in sphagnum moss and shaded by arthritic rhododendrons. It was like wading knee-deep through a huge, engorged sponge, but finally, four hours after leaving camp, I reached the eastern rim of the valley at an elevation of about eleven thousand feet. A cloud rested at ease on a peak to the south, and another cloud had shadowed the Choushuigou. An animal path followed the rim, and I noted the tracks of takin and wild dog; old droppings revealed where a panda had passed. Here at the upper limit of bamboo, stems were only twenty inches tall and grew patchily on meadows among grass and white ever-lasting flowers. At my feet was a deep ravine, its lower slopes almost vertical. Toward the northwest, ragged ridges floated like icebergs on oceans of cloud and rounded hills surfaced like whales. Clouds, sky, and rhododendron leaves shimmered in the sun with waves of blinding light. Over the ridge two hawks slipped sideways, screaming in the wind, their voices thin in the infinite space. Here I felt a sense of liberation; I wanted to shout a message to the hawks.

A few years before I had accompanied several Yanamamo Indians through Amazon rain forest, heading north toward the Neblina highlands that rise on the border between Brazil and Venezuela. After several days the terrain sloped steeply up to where the forest grew stunted and then became shrub and meadow, the views suddenly immense; below us, a sea of trees billowed to the horizon. The Indians had never been there; they had spent their lives in the forest's gloom with no horizon more distant than the wall of trees at the edge of their fields. Each climbed up a boulder and there jumped up and down, waving his arms and whooping with full voice. I knew now how they had felt. As I plunged back into the valley to head toward camp, the magic of the moments on the ridge drained away, but I was energized by the time spent there. And late that evening I noted in my journal, "I saw stars tonight."

On 19 October we finished five days of monitoring Zhen. Hu Jinchu had just returned from leave, and I showed him the results. Instead of a constant low but restless level of activity, her pattern was normal again. Furthermore, she had been on the other side of the ridge today, a quarter mile from the den. Did she move her young? Had she lost it? A youngster is unable to walk well enough to follow its mother until it is about five months of age, and it does not begin to eat bamboo until at least six months, by which time it has twenty-six to twenty-eight teeth. It therefore remains vulnerable to any small predator when left alone; a marten or golden cat could easily enter a den and kill an unattended panda young. I suggested that we check the den site the next day and Hu Jinchu agreed.

20 October. After a dense fog lifts in early afternoon, Hu Jinchu and I push through head-high bamboo into a shallow ravine in the direction of Zhen's den. The radio signal indicates that Zhen is in the area but not at the den itself. Hu Jinchu leads. Suddenly we see bamboo sway thirty feet ahead, and Zhen appears by a fallen tree. She turns to leave, abruptly changes her mind, and, giving two loud roars and several snorts, trots toward us. There are situations when it is unwise to test whether an animal is just bluffing—and this is such a case. I crash through the bamboo to a nearby tree, a small one six inches in diameter, a *Sorbus* I think, and scramble up. Not until I am sitting on a low branch, eight feet above ground, do I look back. Hu Jinchu is retreating somewhat hurriedly back over our route. Confused, Zhen hesitates a moment, first starting to follow Hu Jinchu, then veering toward my tree. After she passes beneath my feet, she halts, huffing and snorting, and listens for one long minute. Everything being quiet, she returns to Hu Jinchu's trail as if intent on tracking him. Instead she turns toward the den tree one hundred and twenty feet downhill. Halfway there, she stops and waits, motionless and silent, as if ready to attack again at the slightest provocation, displaying a sense of lurking violence that is quite unlike her usual benign and complacent demeanor. I can see the triangular entrance to her den but not its dark interior. I hear a loud squawk, then another. Her young is there—and obviously well protected by its irascible mother. As quietly as possible I slip down from my perch and retreat. Farther away, I climb another tree; Zhen has not moved in spite of the noise I have made. Hu Jinchu and I soon meet on the main trail.

On the evening of 23 October we went to bed in autumn; the next morning we awoke to winter. Four inches of snow had fallen, and mountain slopes gleamed now with the muted silver of a panda's back. When toward noon the snow melted and slipped from the boughs with a hissing sound, most leaves came with it, and the ground was soon carpeted with a vibrant pattern of browns and yellows and reds. With winter here, it was time to begin trapping pandas again. We could not open traps within Zhen's range for fear that mother and young might become separated, Zhen inside the trap, her offspring frantic outside; but we hoped to catch another adult female and an adult male elsewhere in the valley. However, no preparations for trapping were being made: trails, partly overgrown with this year's new bamboo, remained uncleared, traps were in disrepair, and bait was unavailable. When I inquired when trapping would begin, the answer was ambiguous. Would we trap at all this winter? Not that I was eager to immobilize pandas; always apprehensive, I was even more concerned now because of recent discussions between China and the Smithsonian Institution.

The Smithsonian Institution had submitted a proposal for a panda study to the Chinese government in December 1978 and had kept this quiet. WWF did not learn of the Smithsonian interest until after it had signed its agreement for such a study with China in September 1979. In May 1980, while WWF and China agreed to the details of their joint project, including the million dollar WWF donation for a research center, S. Dillon Ripley, then secretary of the Smithsonian Institution, continued to discuss a separate panda project with his Chinese contacts. A delegation from the Smithsonian toured three panda reserves in April 1981 and selected Tangjiahe as the site for an intensive study. The memorandum of understanding that the Smithsonian signed that month with the China Association for Science and Technology noted that "it is quite useful to conduct some short-term surveys in certain reserves while longer-term research continues, so as to speed up the comprehensive understanding of the population structure of the Giant Panda in the wild, and thus leading to the adoption of certain measures for conservation and management."

These aims were similar to those expressed in the WWF agreements. Different Chinese departments and foreign institutions were promoting their own panda projects without attempting to coordinate. WWF had made a substantial contribution toward a research center on the basis of China's argument that such a center would be essential for the coordination of all panda research. There was now the possibility of wasted effort and competition between projects, something that did not please me. Such a state of affairs would not benefit the pandas. To promote cooperation, Theodore Reed, then director of the National Zoo, who was intimately involved in the Smithsonian project, generously informed WWF and the New York Zoological Society of the progress of the Smithsonian's plans. The Smithsonian had a team ready to begin field work, including Melvin Sunquist, who had spent time with me in Pakistan, John Seidensticker, and John Eisenberg. During a meeting between the Smithsonian and the Chinese a single question evolved into a major problem. When a member of the Smithsonian party asked about penalties for accidentally killing a panda during immobilization, they were promptly told there would be a fine of five hundred thousand dollars.

Under Chinese law there was a jail term for killing a panda but no fine. The sum was apparently based on the insured value of a panda that had been loaned to Hong Kong for commercial purposes. (However, a Hong Kong dollar was only about one-seventh the value of an American dollar.) Such a demand naturally caused consternation. As a Smithsonian negotiator mildly replied by letter to the Chinese:

> In the absence of applicable laws and regulations, the imposition of any financial obligations due to injury to or death of a giant panda would be contrary to the spirit and intent embodied in the cooperative agreement.

Other options were considered. The Chinese proposed an insurance policy, but the Smithsonian could not accept an admission of liability. The Smithsonian countered with the suggestion that only Chinese veterinarians immobilize pandas. The Chinese, however, noted that this would only reduce liability by fifty percent since it was a joint project. The meeting broke up with the problem unresolved. When the Smithsonian telephoned Beijing to discuss a possible exemption from liability—an exemption that would have to be part of a written agreement—the conversation, as related to me, went like this:

China: "The Ministry of Forestry does not want to set a precedent."
Smithsonian: "The issue cannot be discussed after a problem arises."
China: "World Wildlife Fund does not have such a clause in their agreement. Why should the Smithsonian?"

The Chinese had made an unacceptable demand, and because they made it at an official meeting they were loath to withdraw it. With the Smithsonian wanting a written waiver, and the Chinese not willing to give it, the project died. I suspect that the Chinese side actually no longer wanted it.

I had discussed the dangers of immobilization at two meetings with the Chinese, but the idea of liability in case of death neither occurred to me nor was raised by the Chinese. Should WWF demand an exemption? Should WWF ignore the matter? WWF and the Chinese would have their annual meeting in Chengdu during late October, and my concerns could then be addressed. But whatever the outcome, I would from now on be wary of the cold shifting winds of Chinese attitude in this matter.

Late in October I met the WWF delegation in Chengdu for its annual meeting with the Chinese. I was told informally by the Chinese not to raise the Smithsonian problem at the meeting as it was of no concern to WWF. One problem less. During the meeting we discussed the program for 1982. We pointed out that the action plan specified making "detailed surveys of numbers in sample areas inside all reserves and other panda habitats," but that no work outside the Choushuigou had been allowed so far. Surveys would be done, we were told, but not next year. We suggested that a Smithsonian team come and work under WWF auspices. That would be fine, the Chinese replied, except that it is inconvenient and there are restrictions and problems. However, it was at last agreed that one Smithsonian zoologist, Devra Kleiman, could observe the behavior of captives during the mating season at Yingxionggou and the Chengdu zoo.

On 5 November, I returned to Wuyipeng, this time with Kay, who had arrived in Chengdu two days earlier.

Zhen had left her den area on 24 October and we now wanted to check the site. Qiu Mingjiang, affectionately called Xiao Qiu, and I climbed up the slope. Xiao Qiu, a slight bespectacled, nineteen-year-old who had just graduated from college to become an English teacher, had been sent to

Wuyipeng in late September as interpreter. His English was superb, but even more remarkable was his interest in our research and his enthusiasm for taking part in it. However, excellence is threatening, and Xiao Qiu was in a difficult position as he tried to straddle two worlds, one that did not much appreciate his abilities but to which he belonged, and the other that he was not allowed to embrace.

Although there were no panda tracks in the fresh snow around the den area, we approached cautiously, Zhen's previous fierce defense of her offspring on our minds. Heaped at the den's entrance, like a giant step, was a large pile of droppings, but the nest itself was clean, mostly composed of wood dust clawed from the walls. The floor was about thirty-five inches in diameter, wide enough to permit Zhen to sit comfortably. She had bitten off several birch and fir saplings near the den and dragged them to the entrance, the ends projecting inward; several branches and a log had also been hauled in. And, standing upright in the nest was a little fir, five feet tall. Zhen did not reveal her reasons for collecting all this wood.

As winter settled in during November, we continued our routine of locating the three collared pandas daily and of monitoring them around the clock for five consecutive days each month, the work made pleasant by the first spell of sunny weather since May. The maps on which we recorded the daily locations of each animal were filling up with dots, giving us a good idea of range sizes. Zhen occupied an area of just one and a half square miles. In spring she had made her two-month excursion into umbrella bamboo, but otherwise she stayed near the crest of the ridge that divides the Choushuigou from the Zhuanjinggou. There she settled for months at a time within an area of about seventy-five acres that included everything she needed, ample bamboo, water, and den sites. Although Long and Ning shared a range in the Zhuanjinggou, each also made occasional trips into the Choushuigou. Long's range was two and a third square miles in size and Ning's a little over one and a half square miles. Their ranges overlapped Zhen's at one end, but neither animal spent much time in that area. Instead they had a favored haunt that they shared, but even though they were often near each other they seldom joined: they preferred being alone together. It was a curious relationship, one that I could not explain without knowing their family history. Perhaps Ning was Zhen's daughter, for among large carnivores when a daughter becomes mature she often occupies a range adjacent to that of her mother, whereas a young male may roam widely and settle far from home. Although Long and Ning were the same age, they probably were not siblings, only friends.

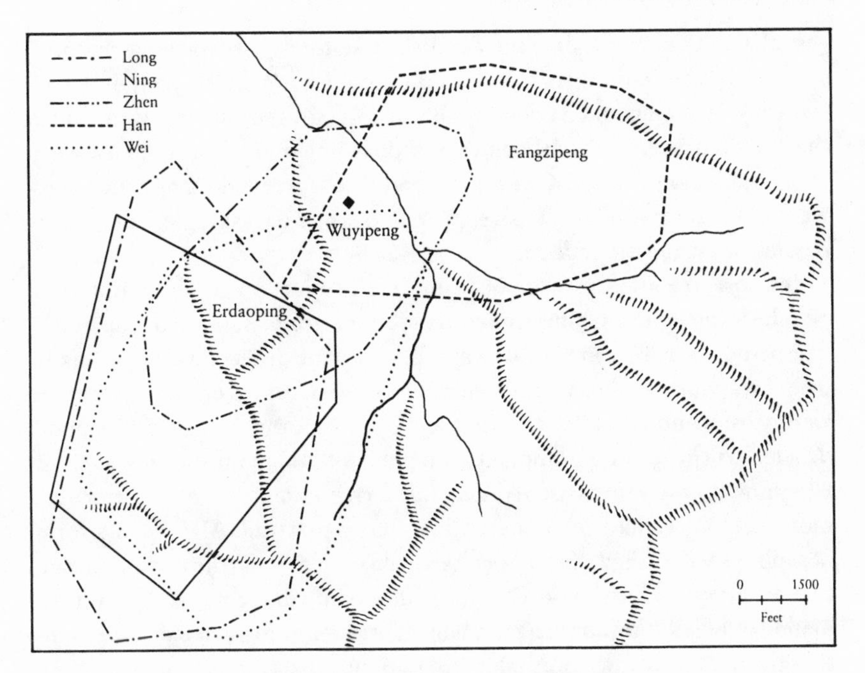

The home ranges of five radio collared pandas

We knew that at least two adult males shared Zhen's range, but we lacked details about their activities and travels; we needed to radio collar and track them. With neither trails cleared nor traps repaired, I offered to hire villagers for these tasks. This suggestion was accepted—except that the camp staff decided to do the work to earn extra money. Several traps were opened and baited in early December, and our daily schedule again included trap checking. I roamed widely, looking for a panda's footprints and also for signs of any other wildlife. One day I flushed two Chinese monal pheasants, large as turkeys, their plumage an iridescent coppery purple. Sometimes I found footprints of monkeys on the trail, as if a horde of tiny barefoot children had preceded me, and occasionally with luck I met the monkeys themselves, usually up in the canopy, leaping like giant golden frogs from branch to branch, landing feet first in a shower of snow that continued to drift like sparkling lace among the trees long after the animals had gone.

On another day, near Fangzipeng, I found wire snares set by poachers in the same place as the previous winter, and also noted fresh stumps where village wood cutters had penetrated far into our study area to fell conifers, casual sins that would in time denude these forests. Displaying the possessiveness that comes with love of an area, I complained to the reserve leaders. Nothing could be done yet, I was told, for the villagers were under the jurisdiction of civil authorities, not the Ministry of Forestry. However, this would soon change. One reason that illegal tree-felling had increased was because everyone wanted a supply of logs and roof shingles before new regulations came into effect.

In work, results are usually in inverse proportion to the number of people doing it. The Chinese maxim of "more workers, more production" hampered our efforts in various ways. But with the onset of winter, so many from camp migrated to warmer climes or were occupied elsewhere that we were actually understaffed for the twenty-four-hour monitoring. Kay and I decided to spend a continuous two nights and a day on the ridge at 4X. Climbing up we were as usual quiet and alert, hoping to see some animal; a month before we had encountered a panda on this trail. At 4X, a platform of saplings with a roof of tarpaper had been built to keep the tiny mountain tent dry. It was a bleak mid-December day, damp and cold, and we immediately crawled into our sleeping bags. Our breath formed ice crystals inside on the tent's low roof; later in lamplight the ceiling twinkled like a night sky filled with stars, and Kay lay there, only nose and eyes exposed, naming constellations—Orion, Big Dipper, Northern Cross, Pleiades. That night the temperature dropped to 16° F, and snow slanted down, its silken curtain isolating us within the circle of light cast by our kerosene lantern. While Kay tried to sleep, the chill seeping into her bag, I took the first shift. Although I had a book to read, it was too cold to bother. Instead I stayed awake by sipping hot water from a thermos, eating peanuts and crackers, and listening to a silence that was absolute.

My boots were frozen stiff when I dressed at dawn to go check distant traps. Wind-blown snow had changed the shape of stumps and knolls and obscured the familiar trail along the ridge. The sun was buried, everything swathed in dense mist. After returning around noon, I pinpointed Long's, Ning's, and Zhen's positions before slipping into the sleeping bag. Kay and I talked little, each absorbed in trying to generate warmth. And when we did say something our voices sounded isolated and muffled by snow and fog. The rest of that day and the night were just as pleasurable, as Kay described in her journal:

> My air mattress kept losing its air so I had to blow it up periodically all night. With no heat, we stayed in sleeping bags all night and day—I kept gloves on, down bootees, down pants, and down jacket over my 3 layers of duotherm top and bot-

tom, long sleeve nylon turtleneck, and wool sweater. I was still cold. Rising on one elbow to tune in to the 3 collared pandas every 15 minutes soon gives one a sore elbow. George and I split the night and I did the day while he checked traps and did azimuths—returning at noon frozen and wet. I was in his sleeping bag to have the better air mattress all day, so he dived into mine taking some of his wet clothing in to dry against his body and putting some underneath the sleeping bag for same reason. He slept part of the afternoon and took over the monitoring about 4 or 4:30 and I switched back to my own bag. We were too cold and miserable to even bother trying to light the temperamental kerosene stove, but our big thermos kept the water we brought surprisingly hot so we had hot tea and cocoa.

Tuesday night I felt awful, too cold all night as sleeping bag was a bit damp from George "drying" clothes in it, my air mattress was sinking regularly, and I was getting the real cold I had been fighting for days. I finally burst into tears at midnight and woke George up after 3 hours of sleep to take over—and my misery made him miserable. But having wept I could then sleep and got 3 hours before time to take over again. After two days of overcast misty weather, which left the trees thick with frost, the sun shone brightly on Wednesday morning and the mountains across from the ridge were beautifully clear. George went off to run the trap lines and I continued monitoring until 9:15 when the 5 days were up—then packed what I could fit into my pack and left for Wuyipeng. Fresh snow on trail and various prints of small animals, but no panda or leopard. . . .

Got home at 11:00, unloaded my snow-covered pack, quickly built a fire in stove, and sat for an hour trying to thaw out, drinking cocoa, and unpacking. Crawled into bed at noon and slept 1½ hours. Fire almost out so rebuilt and back into bed while water heated for sponge bath. George home about 2:30–3:00, tired and cold; I gave him tea and he got into bed and read mail and magazines that had come.

I went to dinner—my first real meal in two days—came back and we both went to bed by 7:30, too tired to stay awake.

December 22 was dazzling, the sky deep blue, the landscape intensely white, the forest filled with bright shafts of sun. Kay and I climbed up the slope behind camp to get a Christmas tree, a unique tree already selected by Zhen. We would bring home the little fir that Zhen had leaned upright against the wall in her den; we thought that tree would give Christmas a special meaning to everyone at Wuyipeng. When less than halfway to the den, we heard insistent hollering from Bai Ai, the word *xiongmao* carrying faintly over the distance. A panda had been caught.

We hurried back to camp, picked up equipment and left for Bai Ai with Zhang Xianti, who had discovered an adult male in the trap there. A Chinese television unit, which was filming the project, had preceded us and surrounded the trap. The panda sat hunched and silent, so drawn into himself that he ignored the needle with which I injected the drug into his shoulder. He was asleep in five and a half minutes. We fastened a collar around his neck and measured him. But just as we started to weigh him, he stood up and with irresistible force pulled away from us, plowing downhill through deep snow, his steps still unsteady. I followed alone and at a distance to make certain he would recover well. After resting briefly in a thicket, he moved steadily up the slope and over a high ridge. From 3 P.M. to dusk I was able to pick up his radio signal—always from that ridge, always inactive. Was he asleep, or had he pulled off the collar? Back in the tent that night, during the long hours before dawn, unable to sleep with worry, I relived the immobilization in my mind. I had slipped my fingers between neck and collar to see that it was neither too loose nor too tight. I felt certain there was no problem. Yet doubts remained when in the morning Kay and I once again climbed up to Zhen's den.

Kay lifted the little fir gently from the hollow. Some of its needles had fallen, there were tooth marks on the trunk, and Zhen had not chosen for symmetry. But we were delighted with our choice. While Kay carried the tree back to camp, I climbed up on the ridge and tuned the receiver to 197, the new male's frequency: he was there in the next valley, the signal rapid, the animal active. All was well.

The next day was Christmas Eve. After we checked traps, Kay and I trimmed the tree in our tent, using simple ornaments we had brought to remind us of other Christmases in distant places. There were paper Santa Clauses, mice made of felt, and other decorations made by our sons when they were small, a wooden chickadee that had perched on a bough of giant heather in the home of the mountain gorilla, and silver cutouts of animals—blackbuck, hare, chital deer, tiger—I had made from the lids of milk tins to decorate a flowering bamboo with clusters of seed that had served as our Christmas tree in a jungle of central India. This year we added miniature pandas and golden monkeys given us by Chinese colleagues. A red wooden star crowned the top. That evening, Kay and I celebrated quietly by ourselves, exchanging books and other small presents, eating fruitcake sent by Hong Kong friends, sipping Cherry Heering, and listening to Beethoven and Mendelssohn on our cassette recorder.

On Christmas Day the leaders of Wolong provided us with a banquet. Two cooks prepared dishes all day. We moved Zhen's tree into the communal shed and placed beneath it small wrapped gifts for our coworkers and for the visitors who arrived from Wolong headquarters and Chengdu for the celebration. Nancy Nash and World Wildlife Fund–Hong Kong had sent sparkling streamers and ribbons and these we strung around the room where they reflected light from the log fire, adding a touch of gaiety to our somber hut. Dong Cai, a new member at camp, had sculpted a snow panda, its eyes and ears blackened with charcoal, and it sat at ease in a large hand basin. There had been many banquets in months past to celebrate someone's coming or leaving and to mark holidays such as the mid-autumn festival in September and Liberation Day on October first, each a congenial occasion well designed to relieve tensions and promote camaraderie, for the Chinese are wonderful hosts. But this Christmas celebration was the most festive event of the year, one that drew us together, all of us remembering our distant families.

The meal began at 6:30 in the evening, with eighteen of us crowded into the small shed as the cooks served course after course, spiced bean curd, steamed egg garnished with red peppers, spareribs, spiced rabbit, meatballs rolled in sesame seeds, fried sugared noodles, pork and vegetable stir-fry, to name a few, all washed down with beer and sweet wine and searing *bai jiu*. Cakes had been sent from Chengdu, and these were happily dismembered with chopsticks. Kay and I received gifts, appropriately a porcelain panda and a panda scroll. And we presented everyone with gifts too: pens, flashlights, socks, and other useful items from America.

Bi Fengzhou, who had come from Chengdu, toasted our project and noted, "The best Christmas present is that another panda has been caught." We enthusiastically drank to that sentiment.

"Tonight we announce the name of the new panda." intoned Hu Jinchu, and paused to stress the importance of his proclamation. "His name is Wei-Wei." It was a fine name, meaning "grand," and we emptied our glasses to Wei-Wei.

Kay followed with a ritual toast: "We usually celebrate Christmas with family and close friends, and this year we are happy to be with our Chinese family." She then led the group in singing "Silent Night," and some joined in with spirit if not musical harmony.

"We have had a good year of cooperation," said Hu Jinchu in a formal toast. "To Peter Scott, Lee Talbot, de Haes, and Halle of World Wildlife Fund, to Miss Nancy in Hong Kong, and to Conway, Dolensek, Quigley, and Stover of the New York Zoological Society, to all these foreign friends far away, and also to the Schaller sons, I send my best Christmas wishes." We all stood up to toast these words. Bottoms up. *Gan bei.*

I followed with a speech noting that we celebrated tonight for three reasons: we had concluded one year of field research in close cooperation, it was Christmas, and we welcomed a new member to our group, Jilin Zhiha.

Jilin Zhiha, a Yi tribesman from the southern part of the panda's range where he is on the staff of the Mabian Reserve, had joined us just three days before. Like several other workers before him from other reserves, he was at Wuyipeng for a few weeks to become familiar with our research techniques. Dressed in a cloak of white homespun and white turban twisted into a horn over his forehead—the site of the soul that no one is permitted to touch—Jilin Zhiha added sartorial elegance to the scene. He readily complied with everyone's boisterous demands to empty his glass, which was always promptly refilled, for he had to sustain the reputation of the Yi as drinkers of epic capacity. Suddenly serious, he rose and announced that he would chant a song of welcome for us in the Yi language, one composed for the occasion:

> With great pleasure I spend Christmas
> with our American friends;
> I am so happy
> that I do not know whether to
> eat or drink.

But the platters were empty of food, and so Jilin Zhiha, to rousing cheers and shouts of *gan bei,* decided on the latter and everyone joined him. The fire burned down and the room became icy, yet faces were red and foreheads shone with perspiration. As more toasts followed us out the door, Kay and I descended the slippery trail to our tent with greater care than usual.

Three days later we caught Ning again. In contrast to her friendly and outgoing behavior eight months earlier, she now cowered, muzzle tucked to her belly in seeming desperation. Her collar still fitted well and would function for at least another year, but I replaced it in the event that we could not catch her next winter. She had gained nineteen pounds since April. When I opened the door of the trap, she walked sedately past me and with barely a rustle pushed through a thicket of snow-bowed bamboo and brambles until she reached a rivulet, where she stopped to drink. Her coat blended so well into the surroundings that she was barely visible.

There has been much speculation about why the panda is so emphatically colored. Some say that the black-and-white pattern is cryptic, helping a panda hide from predators. A panda in snow is indeed difficult to see, as Ning proved, but snow covers the ground only a few months of the year, and at low elevations it may not fall at all. Besides, a panda is perfectly unobtrusive in thick bamboo, no matter what its color. Others say the color may help the panda regulate its body temperature, black losing less heat than white. This may be one function of the coat, but the idea does not explain why the panda has such a clownish face.

Rather than being cryptic, for much of the year the panda's coat is bright, making the animal conspicuous. Striking colors often send complex signals to other members of a species. Pandas generally avoid contact, they are silent except during the mating season, their habitat is dense, and their vision is not acute. I think the bright coat may help pandas prevent close encounters with other pandas. In pandas, as in many animals, a stare represents a threat. The eye patches enlarge the panda's small, dark eyes tenfold, making the stare more potent. In addition, a staring panda often holds its neck low, a position that not only presents the eye patches to an opponent but also outlines the black ears against the white neck, in effect presenting two pairs of threatening eyes. Conversely, to show lack of aggressive intent, a panda averts its head, covers the eye patches with its paws, or hides its face, as Zhen did during mating. If my speculation is correct, then it is curious how in the human mind a threatening image has been rendered cuddly.

A somewhat less evolutionary explanation of "how the panda got its patches" was retold by Nancy Nash in *Young Discoverer, Cathay Pacific* magazine:

> One day many, many years ago a young Chinese girl saw a leopard fighting with a panda. The girl knew that the panda was going to be killed, so she tried to pull the leopard away. The leopard turned on the girl and killed her instead. And the panda escaped.
>
> Later, when the panda found out the girl had been killed trying to save his life, he was stricken with grief. So he called together all the pandas in the world and they held a funeral for the young girl. As was the custom at panda funerals, they all wore black armbands. And they were so sad that they cried and cried. They rubbed their eyes with their black armbands and their eyes became black. To block out the sound of wailing, they held their arms to their ears, and they hugged their bodies in their grief. Soon the black armbands had made black patches on the fur of all the pandas.

At eight on New Year's evening, Kay and I stepped from our tent into the silver night of the snowbound forest. Sounds of laughter and loud chatter came from the hut. Overhead was an arch of blazing stars, and meteors prowled, leaving flashing trails. Against the black slope above us we saw a cluster of five or six stationary lights, small and white and burning brightly. Why, we wondered, on such a cold night, would anyone be on that steep slope with lanterns? Besides, the lights remained in one place. What could they be? Phosphorescent fungi? We looked at them with binoculars and I thought of climbing up to investigate. But like spirits, the lights were suddenly gone and did not reappear then or on subsequent nights. The old year ended and the new one began with mystery.

7 Wei-Wei's World

January to June 1982

On a snowy day about a year after I arrived at Wuyipeng, I followed a panda's tracks along the spine of the high ridge above camp. The tracks veered to the edge of the trail where the animal had obviously turned and backed up to a fir. I examined the tree closely. Within two feet of the base, bark had been rubbed smooth and there was a dark secretion on it; bits of bark had been clawed off above this rubbed area. Bending close to smell it, I could detect a sweetly acidic odor. Suddenly I realized that this was a scent-post on which pandas had repeatedly deposited their personal odor. How totally stupid I felt. For months I had passed this tree without noting the distinctive marks on it. Now I hurried along the entire trail counting scent-posts and found that fully twelve percent of the trees flanking my route served that function, most of them firs with a diameter of six inches or more. Some posts had also been marked with a squirt of urine, much as a dog sprays a fence, leaving a strong musky odor. I knew that pandas of both sexes have a large glandular area around the anal region; and in zoos I had seen pandas rub these glands either back and forth or with a circular motion on stones, walls, and other surfaces. The animal's tail, short and bushy, is usually pressed close to the body, a protective cover for the glandular area. However, when marking, the panda extends its tail and wipes it over objects, like using a brush to paint scent. Sometimes a panda will walk up a vertical surface with its hindpaws and scent mark while doing a handstand. Walking that ridge, I had used my eyes, not my feeble nose. How would I ever understand pandas? They moved from odor to odor, the air filled with important messages where I detected nothing.

An animal like the panda may be solitary, but it nevertheless lives in an organized society of relatives and neighbors, some friendly and some not, depending on previous contacts and personal alliances. Of course, the panda, no matter how misanthropic it may seem, still has two social imperatives: when young it depends on its mother for survival, and when adult it must find a mate. Any society depends on communication, on a system of sights, sounds, and smells. Pandas do not waste time and energy in arranging meetings except when courting; in fact they assiduously avoid contact. One means by which a panda can bypass another is to advertise its presence with vocalizations, and the listener can then decide on a course of action, whether to approach, flee, or ignore the calling animal. Pandas produce calls notable in variety and volume—adults have a repertoire of at least eleven sounds—with which they communicate subtle shifts in emotion. However, they vocalize mainly during courtship and unexpected meetings. Calls have a major drawback in that they are transitory, they must be repeated often to keep information current and to make certain someone is out there to receive the message.

Scent marking, by contrast, is a more efficient way of sending information: the odor serves as a long-lasting calling card that not only identifies the individual after it has passed but also conveys something about its reproductive state and the length of time since it was at that particular site. Pandas, especially male pandas, mark with urine, rub glandular scent, and claw bark. I could see and smell the difference in the methods but had no idea what specific message was sent by each. Most posts were logically located on ridge crests and along other routes much used by pandas, but it was difficult to determine why a post had been abandoned, the rubbed surface overgrown with bright green algae, and a new one created a few feet away, or why a panda in its travels scent marked some posts but not others.

Telemetry is a useful tool in that it helps us to locate specific animals and indicates whether they are active or resting. However, such technology almost turns animals into abstractions, mere points on a map and activity graphs; it reveals little about them as living beings with daily problems and aspirations. I knew that pandas were usually sedentary, moving a straight-line distance of about a quarter mile per day as measured by radio locations on consecutive days. But what did an animal actually do during those hours? Now that I realized the importance of scent in a panda's life, I wanted to follow our newly collared male Wei-Wei, who resided in the same area as Zhen, Long, and Ning.

Before dawn on 17 January, Wei crossed the ridge into the Choushuigou, the shady side of the mountain where snow was quite deep in places. There

I could follow his tracks but would stay one day behind to avoid disturbing his routine.

Wei angled down-slope hidden beneath bamboo, sometimes on a trail or fallen tree but mostly cross-country, walking steadily as if he knew precisely where he was going. He selected his route with blithe disregard of thickets, windfalls, dropoffs, and other obstacles that left me floundering in tangles of stems and hanging suspended from saplings while my feet flailed for support. By trailing Wei at his speed rather than mine, I used my senses more, taking note of the many small things that had been part of his day.

On a rise covered with sphagnum and rhododendron, he scent marked a big, solitary fir, and a little farther on, another. Coming to a steep incline, he slid down on his chest and belly, seemingly from exuberance rather than necessity, and continued parallel to the slope. When he reached a rivulet frozen and covered with snow, he followed it for a short distance to a level spot. There he raked ice and gravel aside with a paw, making a small basin into which water trickled; then he drank. Occasionally he ate briefly. He marked several more trees as if at random, for they were not established scentposts, but I soon noted that he left his odor at places where his route made a sharp turn, as if he wanted to blaze a trail. Twice he rested for at least an hour, and at one of these sites he snacked in bed upon waking. In late afternoon, he settled into a shallow ravine where all night he ate and slept and once drank at a seepage. In the morning he posted his calling card on two firs, and then padded on, along ridges, across logs spanning ravines.

I followed Wei's route for five and a half days. His course took me in two big loops downhill, then up and back where it began. My notes are full of detail. I learned that he averaged just over three-quarters of a mile of travel a day, surprisingly little for such a large animal, that he left his scent on forty-five trees, that he had nine lengthy rest periods, that he deposited an average of ninety-seven droppings per day. But there were many things, to Wei far more important things, that I failed to note or understand, or about which he left no message. What were his feelings when he crossed a trail Zhen had made a week earlier? Why did he suddenly backtrack one hundred and twenty feet? What were his thoughts, if any, on passing a latrine made by his small relative, the red panda? Although I failed to grasp some details, I developed at least a more perceptive mind by following Wei's tracks and could appreciate even more the uniqueness of his personal world.

At best a researcher can reconstruct only a few fragments of an animal's life. However, facts about the panda, even when only partly understood, have importance if they help the species in its struggle for survival. That is why Kay and I were once more on Wei's trail, this time to determine how much bamboo he had eaten in one day. If we determine how much bamboo is available and palatable to pandas around Wuyipeng and how many pounds of leaves and stems an animal like Wei consumes in a day, we can then roughly calculate the number of pandas that these forests could support. Wei had remained almost twenty-four hours in a hollow among leaning birch. The bamboo was bent with snow, and Wei's trails were clearly evident as he meandered this way and that. On the menu were year-old stems and leaves. For hours on this cold and sunless day, we counted, measured, and weighed, and finally calculated that in one day Wei had consumed parts of about twenty-two hundred stems and a portion of the leaves from nearly fourteen hundred stems, or about thirty-one pounds of bamboo. Other pandas on other days sometimes ate less, but average intake was still twenty-five to thirty pounds. This in itself may seem to be a trivial fact and, like any fact divorced from the context of life, a dead one. But as one delves more deeply into its meaning, it bears witness to the panda's wondrous past.

Carnivores and omnivores normally eat meat, fruits, and seeds—all food rich in protein and starches. Such foods are easily broken down and assimilated by the body, and such animals therefore have simple stomachs and relatively short intestines to process their food. By contrast, herbivores subsist on grasses, leaves, and stems, a diet low in nutrients and difficult to digest. As described by Peter Van Soest, plants, when viewed from the standpoint of herbivore digestion, have two main components. One is the cell content, with its soluble nutrients—proteins, starches—and the other is the cell wall, with its tough, fibrous cellulose, hemicelluloses, and lignin. Cell content is easily digestible. However, herbivores lack the enzymes necessary to break down the plant cell wall. To use these parts of the plant, herbivores have formed a symbiotic relationship with bacteria and protozoans that live in their digestive tract and can degrade cellulose and other tough parts by fermentation. But since fermentation takes time, herbivores must retain food in their digestive tract, and this requires a fermentation vat. Horses, elephants, and hares, for example, have both a sac-like cecum and an enlarged gut that store, ferment, and absorb nutrients. Cattle, deer, and other ruminants have a specialized foregut where food is fermented before it passes into the stomach for further digestion.

Now consider the panda. The species has retained the simple digestive tract of a carnivore: it lacks a special chamber to retain food, its intestine is short, and it has no symbiotic microbes to ferment cellulose. Unable to extract nutrients from cellulose, it must depend mainly on cell content for needed energy. Bamboo is a diet of exceptionally low quality for a panda. Bamboo consists mostly of cell wall of which cellulose and lignin are indigestible and hemicelluloses only partly digestible, as shown by our research as well as that at the Calgary Zoo and the National Zoo in Washington D.C. A panda can assimilate only about seventeen percent of the bamboo it eats, an extremely low figure. Even a goose, renowned for the rate of its food passage, has an efficiency of about thirty percent, and deer subsisting on green grass may achieve up to eighty percent. Meat-eaters like the lion have an efficiency of ninety percent. One fact is clear: because of its inefficient digestive system, much of what the panda eats is useless bulk.

Everything an animal does requires energy—digesting food, pumping blood, keeping warm, scratching its head, feeding on bamboo, and any other activities needed for basic maintenance. In addition, energy is required for growth, whether the animal's own or that of a fetus. Balanced against these demands is the energy obtained by an animal from its food. Although Wei spent over thirteen hours a day chomping stems and stuffing leaves to keep his stomach filled, he obtained few nutrients in return. I calculated that from all those pounds of bamboo it eats, a panda's daily intake was only about forty-three hundred to fifty-five hundred kilocalories per day. And energy expenditure, based on the panda's basal metabolism and caloric costs of feeding, traveling, and other activities, may be four thousand kilocalories or more per day, or only a little less than the intake. This means that the panda's nutritional margin of safety is fairly small, that it has no extra energy to waste. One could logically argue that if the panda needed more energy, it could simply devote more time to eating. But this it cannot do because its stomach capacity is limited and digestion takes time.

Such a low-nutrient diet has affected many facets of the panda's life. A panda cannot eat enough bamboo to accumulate much fat, and therefore it cannot store sufficient energy to survive several months of hibernation. While a bear sleeps the winter away, curled up safely in a hollow, a panda must continue to eat in snow storms and bitter cold. In fact, the panda can survive on bamboo only because this plant has several unusual features. Bamboo belongs to the grass family, and grasses in areas with marked seasonal changes in climate, whether cold or drought, fluctuate in nutrient levels; for example, the dead, dry grass eaten by elk in a Wyoming winter has much less nutritional value than the green grass of summer. Bamboo is not only abundant and green all year but also it maintains a more or less constant nutritional level. There is no seasonal feast or famine for the panda.

With its bare subsistence diet, a panda must keep its digestive tract filled to assure itself a steady flow of calories. Instead of sleeping for long stretches, it must forage at intervals throughout the day and night. It must also have an efficient feeding strategy—selecting and eating without wasted motion, using its special paws and teeth for processing. The panda's need to conserve energy permits only slight flexibility in behavior. Anything requiring a burst of energy is avoided or reduced to a necessary minimum. This includes climbing up very steep slopes, ascending trees, and running. Travels are circumscribed, and males like Wei do not waste effort on territorial defense; social interactions, which can be costly in terms of energy, are avoided. A panda mostly eats and rests. A female is even frugal in what she invests in reproduction. She does not conceive until the advanced age of at least five and a half years, the actual gestation period is unusually short, the newborn is tiny, and only one young is raised. If the young dies soon after birth, the energy drain on the female has been small. If it survives, independence from the mother comes early, usually by twenty-two months.

I have speculated about how bamboo nutrient content, panda digestive physiology, and panda energetics have interacted to influence most aspects of the panda's life. A bamboo diet obviously must have presented the species with some basic advantages during evolution. Indeed the giant panda may have become a giant as an adaptation to a bamboo stem diet, large body-size giving the animal the physical strength to harvest and eat tough stems. But why did pandas tie their fate to bamboo? There are, of course, advantages to having superabundant food available at all seasons, even if that food is of low quality. Yet I find it inexplicable that pandas usually ignore the many nutritious plants that grow in the forest during summer, plants upon which black bears feed and grow fat, and some of which even pandas sample on occasion. Are there other advantages to a bamboo diet? Our scientific inquiry has so far provided no satisfactory answers.

A diet of thirty pounds of bamboo a day. This represents not just a fact or even a life-style but the core of Wei's existence, his one great vision in life.

Or at least I think it does.

As an explorer in the ways of animals, I carefully observe what an individual does and write down everything to capture what I have seen. Later I take the unadorned facts and impersonal descriptions, neatly tabulate and graph them, and express the information logically, as if the work had progressed step by step to an inevitable conclusion. Truth and science are served. Lacking in such reports is the human factor, the joy of discovery, the pleasure of a new insight, the admission that research is sporadic and haphazard—and the fact that the information is not as objective as one would like to think. Statistics may help to describe the universe but not other beings; numbers cannot convey the quality of a creature, they cannot express love, anger, joy, and courage.

In addition, as Werner Heisenberg noted: "We have to remember that what we observe is not nature herself, but nature exposed to our method of questioning." A fact is not a fact until someone has posed the question, and slowly the world of an animal emerges from the questions raised and facts collected. But if someone else asks a different question, a different creature, a different reality, comes into existence. The animal is an illusion created out of the animal's interaction with an observer who decides what to measure and record and what to ignore. We constantly infer the unseen, we confuse ideas with facts. Furthermore, animals bound, prowl, slither, and flap through our subconscious in the form of myths about lions, bats, foxes, owls, snakes and doves, each culture with its own fantasies. Victor Hugo wrote, "Animals are nothing but the forms of our virtues and vices, wandering before our eyes, the visible phantoms of our souls." In such a way is science fashioned. Any biologist who observes a tiger, gorilla, panda, or other creature and says he or she has done so with total objectivity is ignorant, dishonest, or foolish.

Such thoughts were in my mind that winter because Hu Jinchu and I would leave Wolong in June for several months to write a report on our panda research. The task would be difficult. We had so far collected much valid and useful information. How would our two cultures, our two different systems of thought, treat this common subject? The Chinese biologists with whom I had talked and whose scientific papers I had read used methods, discussed results, and provided insights that seemed wholly Western in concept without even, as the Chinese would say, "eating a Western meal in the Chinese manner." It would be rewarding to find out what Chinese perspective Hu Jinchu had brought to the study of pandas.

The problem of penetrating the panda's mystery was made even clearer to me one foggy afternoon that found Hu Jinchu and me in an almost impenetrable bamboo thicket. Suddenly the fog grew intensely bright as if sundered by the sun, and in a shaft of strange light a panda was before us, leaning eerily calm against a fir, paws resting in his lap. At first I thought it was Wei, but he was uncollared. His expression was stern but with a touch of humor, and his eyes were vivid like a hawk's. I was somewhat astonished when he abruptly began a monologue that seemed to transcend ordinary language—though perhaps the magic of the occasion was only in my mind:

Honorable scientists: I want to compliment you on your efforts to study my kind. It takes dedication of the highest order to measure so many droppings. Day after day you follow my tracks with admirable persistence if not technique: I can hear and smell you from far away. Actually I'm not certain what you expect to gain from invading my privacy. You generate numbing statistics about the number of stems I eat in a day and the number of hours I sleep. Remember, time is not the same for all living things. This fir lives more slowly than you, and I more quickly. And as to counting stems, it merely shows that you have discovered some easy facts about me; most aspects of my life cannot be written in the language of mathematics. How can you understand me? We may seem to share certain moods, but you cannot comprehend mine. After all, it's not *your* perception of reality that matters. Look at each other. Your ways of thinking are vastly different, yet you belong to the same species.

You people are unbelievably conceited. Just because yours is the most prominent intelligence around, you also act as if it's the highest. There are many serious problems in your style of thinking; you must overcome your ideological prejudices and other unhealthy scientific tendencies. Some of you, for instance, hold that language is necessary before one can think, and that makes me and all others—except you—unthinking creatures. What frivolous nonsense! What arrogance! Your laboratory people first look to see what an animal can't do and then attempt to teach it that. Study what an animal can do best. No wonder someone even referred to me, a panda, as an interesting example of color over content. How can you measure my intelligence, my way of thinking, by your standards? At least you two merely try to observe, not judge. Keep in mind that we live in different worlds, each unimaginably different, I primarily in a world of odors, you of sight; I think mainly with smells, you with words. Ludwig Wittgenstein wrote, "If a lion could talk, we could not understand him." An intelligent remark for a human, and a foreign barbarian at that.

The panda shone with a blinding white light, and his body became dim and shapeless as smoke and dispersed. Just as casually it condensed again into a palpable yet transitory creature, form and emptiness coexisting, matter and space inseparable, energy and spirit integrated.

Another point. You study my diet, you study how many times I scent mark and mate and how far I travel. Remember, you cannot divide me into independent fragments of existence. At best you might perceive an approximation of a panda, not the reality of one. I am, like any other being, infinite in complexity, indivisible, a harmonious whole.

Let me make some suggestions for improving your work. Forget science now and then; silence the rational mind. Try to make my life real by using intuition, sensibility, and empathy. Never forget that all living things are part of the bamboo of life. All animals, even humans, are one great living being. You cannot describe a panda without also describing yourselves; separateness is an illusion. Send me your philosophers, poets, and children, for they will see me with a new vision. Even so, we shall always remain of two worlds. Humans can never know the truth about pandas. Therefore, enjoy the mystery—and help us endure.

With that emphatic ending, the panda pulled the fog around himself, the glow of light contracting to a spark that died.

But he left two droppings by the tree, and these I collected.

A biologist must have in mind some questions worth answering before beginning a field study, some valid goal to fulfill expectations and justify the funds—usually public funds—spent on the quest. I had asked, "How is the giant panda adapted to bamboo?" thinking that the results would not only be scientifically intriguing but also of value to conservation. We learned that bamboo is a low-quality food, so low in nutrition that in spite of the giant panda's stuffing itself for many hours a day, its daily caloric intake barely exceeds expenditures. And we concluded, among other things, that the giant panda's large size seemed to be an adaptation to this poor diet, because physiological and other reasons make it possible for a large animal to use its food resources more efficiently than a small one. I naturally thought that we applied impeccable logic to derive our conclusions. However, the red pandas in the Choushuigou made me uneasily aware that something might be lacking in our assumptions.

In their love affair with the giant panda, people often remain either unaware of or forget that there is another panda, the red or lesser panda, a creature not in any way "lesser" except in size, it being one of the most attractive and appealing of all mammals. Take a giant panda and reduce it drastically in size, change the white color except on the face to a rich rust-red, add a long ringed tail like a raccoon's, and you have a red panda. Its blunt, broad face with its black mask, and its chunky body with stocky black legs resemble those of the giant panda, only the whole bundle weighs twelve instead of two hundred pounds. The red panda has a wider distribution than the giant panda, occurring mainly in temperate forests from Nepal eastward along the Himalaya through northern Burma into China, where it often shares the forests with its larger relative.

When I began research in Wolong, the most informative report of the red panda's habits in the wild was still one published by Brian Hodgson back in 1847, and I was eager to learn more about the species. Red pandas, however, were rare in our Wolong study area; I saw only one, dashing across a path. But I found their droppings—as if made by a miniature giant panda—and followed the animals' snow trails as they meandered through the bamboo. These efforts showed that, unlike giant pandas, red pandas had a small home base at which they had favorite resting spots, a latrine or two, and several scentposts, among them logs dribbled with urine. In vain I listened for a red panda's call. "The Chinese call it 'child of the mountain,'" noted Père David, "because its voice imitates that of a child."

It was clear that red pandas ate bamboo almost exclusively, but unlike the giant panda they took no stems, only leaves and shoots in season. In addition, they ate wild cherries, gooseberries, and a few other fruits in autumn, as Don Reid discovered. If a giant panda can barely subsist on bamboo, how can the red panda manage it? Because of its small size, it needs disproportionately more calories per day. Were our conclusions regarding giant pandas incomplete or wrong? To compare the daily energy requirements of red and giant pandas, I would also need to investigate the red panda's activity cycles and movement patterns, something possible only with telemetry. However, the action plan did not include a study of red pandas.

One February day later in the project, a red panda was fortuitously caught in a giant panda log trap. Just as fortuitously, we happened to have a suitable red panda collar handy, the transmitter no larger than a matchbox. Ken Johnson, Alan Taylor, and I had recently arrived at Wuyipeng; joined by two Chinese assistants, we climbed up over the ridge into the Zhuanjinggou to the trapped animal. It was a crisp day with clouds settled in the valley except for flamelike tendrils creeping up into the high ravines. When I pulled the panda from the trap, her neck loosely held in a noose at the end of a short, specially designed catching pole, she was understandably furious, giving sharp little coughs almost like a spitting cat. With needle claws she raked Ken as he tried to hold her so we could fasten the collar. She was a dainty female, probably subadult, weighing only eight pounds. Minutes later she was free again. For nine months, until the transmitter ceased to function in November, Ken and the Wuyipeng staff monitored the little female who was named Jiao-Jiao, meaning "First."

Like a giant panda, Jiao was solitary. However, she was once in a bamboo patch with another red panda, the two resting thirty feet apart, not exactly overcome with passion for a close encounter. Monitoring showed that Jiao was active for only nine hours a day, as compared to an average of fifteen hours for giant pandas, and that she was mostly nocturnal, moving infrequently between ten in the morning and seven in the evening. We reasoned that Jiao needed less feeding time than a giant panda because she processed her food more carefully. Instead of taking fistfuls of leaves, including dead ones, as giant pandas are wont to do, she delicately ate leaf by leaf, choosing the most nutritious green ones. And she did not bolt her food but chewed meticulously, breaking down plant tissues so that they could be more efficiently digested. Jiao's home range was surprisingly large—about one and a half square miles—almost as large as Wei's if his seasonal excursion into umbrella bamboo for shoots is excluded. At least five giant pandas, including Wei, shared the area with Jiao. Wei and Jiao even favored the same patches of bamboo and quite often were near each other, but we have no idea how they reacted on meeting. We were even more astonished to note that Wei and Jiao averaged about the same amount of travel each day. Surely tiny Jiao had no need to move as far as bulky Wei just to eat bamboo or for any other reason. We had expected Jiao to be highly sedentary in order to conserve energy. Was she atypical? Jiao picked up extra calories by selecting her food carefully and chewing it well, but she expended these calories in travel. Or so it seemed. By the time the transmitter died, we had discovered interesting new facts about a red panda, but we still did not know how she could meet her energy demands.

Don Reid later radio collared two other red pandas in the same area where Jiao lived, an adult male and an adult female. Their ranges were much smaller than Jiao's, less than half a square mile each. They were more active in daytime than at night, the converse of Jiao's behavior, and they spent about eleven hours a day moving around rather than just nine hours. In a study conducted by Pralad Yonzon in Nepal's Langtang National Park, female red pandas also had ranges of about half a square mile, whereas some males had ranges of up to three and a half square miles. Though such differences in behavior are interesting, they do not help to explain the red pandas' ability to balance caloric intake and energy expenditure.

The solution to the riddle came from the laboratory, the only place where it could have been found. Brian McNab of the University of Florida has for years done fascinating physiological research on animal energetics. While we were working on pandas in the wild, he discovered that the red panda can greatly lower its metabolic rate in cold temperatures by reducing its peripheral circulation, thereby lowering skin temperature and heat loss to the environment without affecting its temperature in the vital interior of the body. A red panda's metabolism at rest was reduced by nearly half, a tremendous saving in energy achieved by turning part of the body's thermostat down. Such a reduction in metabolism must be brief and interrupted by periods of activity because gangrene in tissues may occur if blood flow is too restricted for too long. This metabolic strategy to conserve energy corresponds well with what we observed of Jiao's almost reptilian response to cold; no wonder she seemed stiff and lethargic when on several occasions we inadvertently awakened her from a rest. Thus, red pandas are even more specialized on bamboo and may have an even more precarious energy balance than giant pandas.

We were a small group at Wuyipeng that January of 1982, hardworking, the cold drawing us together and isolating us from casual intrusion. The usual staff was gone, leaving the camp with three temporary helpers and Xiao Qiu, the interpreter. Jilin Zhiha, the Yi tribesman, was the most conspicuous presence in any gathering.

The Yi, or Lolo as they were once called, have a reputation as fierce fighters who in the past killed Chinese and an occasional Westerner who intruded on their terrain in southern Sichuan and Yunnan. Jilin Zhiha was treated by the others with the noisy deference and rough bantering tone that revealed a combination of unease and attraction—not unlike that shown me at the beginning. Jilin Zhiha, to live up to his people's reputation, told tales of murder and mayhem over and over.

We were also busy checking traps during January. Both Wei and Ning were caught and quickly released. Long, too, was tempted by the meat. Now a husky teenager, weighing thirty-four pounds more than the previous March, he was returned to freedom with a new collar. Across the valley in Fangzipeng, we trapped a female whom we named Han-Han. She was so large that at first everyone assumed she was a male; and she was beautiful, the most perfect panda I had seen, harmonious in proportions, radiant in her luminous coat. A few days later we found a subadult male, only two and a half years old, in another trap. Unlike gentle Long and Ning at that age, he showed his abundant displeasure at being trapped, squealing and roaring and striking out at anyone peering at him. Someone asked me, "Do you want to put a collar on such a small animal, or wait for a big male? The action plan says you are only allowed three pandas a season, and we have already done Wei-Wei and Han-Han." An interesting question. Especially since the action plan did not specify a limit and the subject had never been discussed. Someone in the bureaucracy was trying to limit the amount of data we could collect. One learns cynicism the hard way in China. There was no point in asserting myself by waving the action plan; confrontation gets one nowhere. Trapping would still stop after three pandas on some pretext such as the unavailability of bait. I ignored the issue, put a tag in the young male's ear so that we could recognize him—yellow tag No. 81—and released him without a collar.

We often checked Zhen's tracks and searched areas where she had been. Neither then nor later was there any sign of her young. She had somehow lost it.

When not helping in the field, Kay remained busy at camp. Her journal describes a typical day:

> Camp routine is reasonably settled now that traplines are set up. George gets up at first light to check one of the traplines. I snuggle down a bit longer, then dress partly in the sleeping bag to keep warm. If we forgot to empty the wash basin at night, I break the ice and dump it. Pour in warm water from the thermos and take a quick sponge bath. Then up to the shed for breakfast, drinking my tea sitting by the smoky fire. Then a bowl of rice gruel and a cold hard-boiled egg. A second cup of tea. Back to our tent to build a fire in the stove. Once it is going well I leave to go check the leopard trap. I enjoy my hike to the trap each day. [We never caught a leopard.] Yesterday's mist left all the trees and hanging lichens and mosses looking ghostlike with frost.
>
> When I return to the tent I check the fire and drying oven to see if any panda droppings are ready to sort leaf from stem, collect more firewood to dry by stove, write in my journal, write letters, read, heat water for the thermos or for laundry. I am always doing laundry—it is easier to do a small amount each day as it must usually be hung in the tent to finish drying. And living with our smoking stove or the smoking fire in the shed, we and all our clothes become sooty. I boil the bath towels but they still look grimy. After a lunch of noodles, I either do more oven-drying of specimens, teach English to our co-workers, or study Chinese.
>
> About 4 o'clock the panda snare lines must be checked again, the later the better. Today I went with George—2 hours of walking along a narrow trail on the mountain side. It is more a ledge than a trail, though in some places it has been widened a few inches by a sapling along the outer edge. Up and down, up and down; some loose shale ledges slide out under your boot. The last group of snares is down along a creek, and after checking we must climb up a very steep trail—more an animal path to my mind and not always distinct to me—like going up a steep stairway. In this one climb we regain almost all the altitude lost and my legs tremble at the end.
>
> Then dinner, talk around the fire, back to the tent to read, write, or just go to bed.

Spring Festival, or Chinese New Year, began that year in late January with the new moon. It is a time of festivities when everyone tries to be home with family and friends, when special foods are prepared, and firecrackers are set off to scare away evil spirits. The Chinese base their calendar on the ancient lunar system. The first New Year celebration began when Buddha summoned all the animals in creation to join him for a feast. Only the rat, chicken, tiger, snake, dragon, horse and six others came, and to reward the dozen, Buddha named a year in honor of each, to be repeated in twelve-year cycles. This would be the Year of the Dog. On the evening of 24 January, there was a big banquet down at headquarters. An important dish was dumplings filled with meat, symbolic of silver ingots. Without such dumplings at Spring Festival, you will not make your fortune during the coming year. The next morning Zhao Changgui, Wolong's leader, invited Kay, myself, and Xiao Qiu to his home for the traditional New Year's breakfast of *mahua,* deep-fried braided bread sticks, and *tangyuan,* sugar-filled dumplings of rice flour. Zhao Changgui, quiet and with a courtly politeness, was always helpful and concerned for our welfare; soon after the New Year he was transferred elsewhere, and we were sorry to see him depart. That night at camp there was another banquet, followed next day by yet another, making it the traditional three in a row, a surfeit of food, wine, and toasts.

February was a difficult month, for everyone at camp suffered from the Wuyipeng equivalent of cabin fever, the staff like dead trees and cold ashes, as the Chinese would say, the unremitting hard winter work having created weariness. I looked forward to March, when koklass pheasants would again crow at dawn and primulas would bloom, announcing the coming of spring. However, that month would also mean meetings in Chengdu and an invasion of Westerners as the artificial insemination program resumed at Yingxionggou, a WWF botanist began a bamboo study, and a television team filmed the project.

WWF and the Ministry of Forestry thought that close cooperation between the Chengdu Zoo and Yingxionggou would be of mutual benefit in our efforts to breed pandas. The Chengdu Zoo was unfortunately managed at the time by a committee of three persons whose bond was a cordial dislike for each other. I dreaded my semiannual meetings with this committee. Indecision was its key to flexibility, and it was neither inhibited by past agreements nor easily swayed by a spirit of cooperation. The previous year, for example, after we had agreed to collaborate on an artificial insemination program, the zoo quietly inseminated two females while Emil Dolensek and Stephen Seager waited at the hotel for the work to begin. Last year, too, it was agreed that Janet Stover would assist in hand-rearing newborn pandas and that Devra Kleiman would observe the behavior of pandas at the zoo. Now at this March meeting the committee refused to cooperate—but not until after it had accepted several expensive items of equipment that had been demanded of WWF as a basis for collaboration. Emil, who had brought the equipment, offered to help assemble and install it, but his offer was brusquely declined. The zoo promptly burned out the electrical converters, making the equipment unusable. Meetings with the zoo ceased.

Julian Campbell, a British botanist, arrived in Wolong with little more than a plant press, a gadget for measuring soil moisture, and a bag of beans, which, being a vegetarian, he planned to cook for himself. The Chinese, with their appetite for equipment, had expected much more. They puzzled how Julian, with his meager scientific kit, would study such a complex subject as bamboo ecology, and they were not much enlightened when he scattered beer bottles over the forest to serve as rain gauges. Julian had a detached air that camouflaged a persistent nature. He planned to work here for years, he told me, and become known as "Chinese" Campbell, following the footsteps of "Chinese" Wilson, who had botanized here decades earlier. He also planned to distribute handbills with three messages to the villagers: do not cut trees because it causes floods; have only one child or none; and during sexual intercourse for pleasure use contraceptives. While the sentiments were appropriate, the gesture would not be appreciated, especially by the Qiang, who like other minorities are exempt from China's one child per family rule. Julian was persuaded to concentrate on bamboo reproduction instead. But this plan also failed. The Chinese and Julian had different perspectives on how to cooperate and he left within a few months.

A television team from the National Geographic Society and WQED in Pittsburgh arrived with a ton of gear, not only film equipment but also such delicacies as Granola bars, salami, cheese, and peanut butter. The crew, consisting of producer Miriam Birch, cameraman Norris Brock, and soundman David Clark, had a tent at Wuyipeng for use during their intermittent stays there. It became a welcome social club for Kay and me, not because of the exotic foods and Japanese Kirin beer, though we appreciated these, but because of the companionship offered. Norris told us about his raft voyage on one of Thor Heyerdahl's expeditions, and David told of a trip down the Yukon River. Miriam and Kay happily discussed musicals and theater. And we talked over possible film sequences. They were concerned about problems: The Chengdu Zoo wanted one hundred dollars per minute for filming rights; Yingxionggou hesitated to let pandas into the outdoor enclosure for fear they might escape. We chatted, confided, expressed ideas candidly without worry of being reported; it was a verbal feast.

During the day the soft whirr of Norris' Arriflex camera became part of our activities. Certain phrases became as insistent as the calls of a brain-fever bird. "Let's just do that one more time." "Now do me a favor. I'm going to try something different. Could you. . . ?" The forest heard strange jargon: "Show me the slate." "Eighty-six head."

Always on the prowl for a sequence, Norris asked of no one in particular: "What are you going to do now?"

"I'm going to wash my hair," Kay offered facetiously.

"Good, let's shoot it," replied Norris, much to Kay's consternation.

One evening when Kay and I were in the tent on the ridge, ready to spend the night monitoring, Norris and David appeared to film that aspect of our work. As a surprise, they brought cheese and a bottle of Chardonnay, and we had a small party there in the fog and gathering dusk, Wei, Zhen, and the others supplying our music with the beep-beep-beeps of their transmitters.

On 4 April, I found an anemone in bloom and on 15 April a rock squirrel came to camp, its first appearance since the start of its hibernation in mid-November. *Tragopan* males, sporting crests of orange-red feathers like flaming toupées, courted with the screams of a distressed infant, living up to the Chinese name *wawa ji,* baby chicken. And a bamboo shoot was up on 17 April. Spring had returned. But it was a dismal spring, with wind-driven sleet and rain that pelted early flowers into submission. Wuyipeng looked like an Alaskan gold-rush camp of the 1890s, muddy trails, sagging tents, garbage all around. Twenty-five people, including Chinese and American photographers, crowded into eight tents and the hut. Many had colds. I think it was David who observed, "If we were filming a musical rather than a documentary, we could call it *The Sound of Mucus.*"

Everyone was waiting for a repeat of Zhen's 1981 courtship performance. But she wisely refrained from advertising this year's tryst. And when she was near Wei and another male, Norris and David were at Yingxionggou. Even though bamboo shoots were available, Zhen did not descend to the lower slopes until 15 May, two weeks later than the year before. Both that day and the next I took Norris so close to Zhen that we could hear her peel and crunch shoots, but she preferred to remain invisible. Tomorrow we would surely capture her on film, or the day after. But the crew abruptly packed up and left Wolong on 17 May. And, of course,

Zhen then became cooperative; she and I had several long and memorable meetings during the next two weeks. Although David and Norris had waited quietly near Zhen for hours, listening, hoping, until chilled through by rain and glacial fog, they had filmed her only once, a brief glimpse of her head, the only sequence of a panda in the wild they obtained.

In filming as in everything else concerning pandas, there was competition to be first. For a fee of two hundred thousand dollars, ABC television had been permitted to film southeast of Wolong in 1981, and the crew had also returned briefly in 1982. The team included several experienced wildlife photographers, and they obtained the best sequences of pandas in the wild during the 1980s. ABC and National Geographic-WQED, each wanting to promote its program, played a little one-upmanship, as this excerpt from the magazine *Sports Illustrated* (14 June 1982) suggests:

> "We were blown away when we discovered that ABC also had these rights," said Miriam Birch, *Geographic*'s producer, who had just returned from a trip that she had hoped would provide the first Western TV footage of pandas in their bamboo jungles. . . .
>
> "ABC is using Roger Caras, who wrote the GEO piece, as their host," she added, referring to the ABC naturalist, whose August 1981 article on pandas was illustrated by photos that showed supposedly wild pandas that were actually in a pen.
>
> . . . Caras fumed last week that he has no intention of allowing the *National Geographic* crew to impugn *his* network's panda expedition. "I don't think *National Geographic* got any wild pandas on film. *We* got five." he said.

Also in the sweepstakes was a Chinese television unit and later, in 1984, a crew from Japan.

With perhaps two million dollars spent on these four panda films, what was the result? The ABC documentary had some good wild panda footage, but the editing and narration lacked focus, and the film was inappropriately shown as part of a program called *American Sportsman*. (A later remake was far superior.) The *National Geographic* special emphasized captives and was surprisingly uninformative, given all the available data. The Chinese film best portrayed life at camp and the beauty of dripping boughs and cascading rivulets. And the Japanese documentary concentrated on a panda that scavenged at Wuyipeng. The panda still awaits a dedicated photographer to pass on the memory of its lonely life and gallant defiance of fate.

Zhen had mated with a male larger than Wei in 1981. Would Wei be a successful suitor in 1982? With voyeuristic persistence we radio located

Wei and Zhen daily during April and shadowed them, hoping to hear the discordant concert that indicated a panda get-together, but with little success. There were three adult females, including Zhen, within Wei's range of two and a half square miles, a rather small selection when shared among several males. At least Wei tried. Once he followed Zhen for days, but always at a distance, like a bashful and infatuated teenager, almost meeting her but not achieving true contact. He first approached her on 30 April, not far from Wuyipeng. That night she walked up and over the ridge. Wei followed the next day to linger near her, and he also tagged along when she crossed and recrossed the ridge on succeeding days, apparently following her tracks by scent. On 7 May we listened as Wei and Zhen had a quarrelsome hour together in dense bamboo.

The next morning, while Zhen rested, Wei was several hundred feet away harassing a subadult panda. It was a small animal like No. 81 but without a tag in its ear. The youngster was in a shrubby tree, and standing on a swaying branch below was Wei. With one forepaw, Wei gripped the tree and with the other he swatted, but the small panda slapped back just as vigorously. Both were squealing and barking. They then relaxed somewhat, Wei at the base, the other at ease on a branch. After at least five hours in the tree, the youngster suddenly descended past Wei and left. The next day, Wei and Zhen parted too, if their loose association qualifies as being together, his social life apparently over for another year.

Would Wei ever meet a congenial consort? Wei was back on the courtship scene as persistent as ever the following year, in 1983, and at that time he provided me with a partial answer to his fate.

15 April. High on the ridge in fog and rain, I meet two of the Chinese researchers hurrying to Wuyipeng to let us know that pandas are courting on a ridge dividing the Zhuanjinggou and Yingxionggou. A female, they tell me, was up in a fir while Wei hovered around the base. Another male, large, powerful, and uncollared, with blood on forehead and ear, approached and chased Wei off. Undeterred, Wei returned and climbed a few feet up the tree, the female hooting at him, but with the arrival of two males at the base he descended. One was the large uncollared male, the other was that male's equal in size, a two hundred and thirty-five pounder named Pi-Pi that we had collared the previous December. *Pi,* like *mo,* is an ancient Chinese name for the panda; *pi* means "brave." The two giants faced each other, roaring and tussling, until the uncollared male withdrew. Pi then ascended the tree and mated, both animals balanced precariously among the branches. After about an hour, Pi left the female and departed.

I hurry to the scene. Pandas squabble in bamboo downslope. As I push through the undergrowth, I spy a panda walking away only a few feet in front of me. On climbing a stump for a better view, I see a second male lying on his back between the roots of a fir as if lounging in a garden chair. Several sudden loud roars startle me. There, just ahead, the big uncollared male walks around the female, guarding her, chomping his teeth. A branch snaps under my foot, and the male, alert to any competitor, snorts and steps toward me. After mating, Pi apparently relinquished rights to this female. Five males are together, a remarkable aggregation. Two are large and three medium-sized, two of the latter somewhat smaller than Wei. Long, who at three and a half years is not yet adult, is unconcerned in the valley nearby.

With clothes soaked by wet bamboo, I am soon chilled, and after an hour's watching, when all is quiet, I hike the two hours back to camp.

16 April. It snows all day, and a dense fog swirls in the valley. Large wet flakes settle on me and the surrounding bamboo as I watch the pandas in a logged ravine on the opposite slope. The bamboo around the base of a lone fir is trampled, and at the top of that tree the female crouches. Pi appears and trots downhill toward the other large male, who backs away. Making contact, the males growl, roar, swat, and spar, mouths wide open, trying to grab and rip with their massive teeth. The large male breaks away and hurries to the fir, which he scent marks. In the meantime, Pi circles uphill and again confronts his rival, his strength and momentum driving the other downhill; when the latter turns, Pi bites him in the shoulder. Once both rear up and grasp each other in a fierce dance, the valley filled with their screaming roars. During a scuffle, the large male slides over a precipice to tumble twenty feet, but with aggressive determination immediately returns to the battle.

Silence. The sudden transition from uproar to calm is eerie, like the eye of a hurricane. Pi moves out of sight, a natural force with its energy spent; the other male forages briefly, and then he too disappears from view.

Suddenly the two medium-sized males appear, one chasing the other downhill. Inattentive, the one in the lead tumbles squealing over the same precipice that had earlier claimed the large male, his body making a complete mid-air turn before landing with a thump.

The female descends from her tree. Neither Pi nor the other large male responds to her availability. Apparently they fought so savagely not for the female, with whom one or both had mated yesterday, but to assert status. By successfully demonstrating his strength, Pi maintains his position as top panda, and this status gives him priority of access to any other female in heat. Poor Wei. Already middle-aged, he will never grow to the size of Pi and achieve high rank.

But now something curious happens. A medium-sized male goes to the female and mounts her; she merely sits, ignoring him, until he pushes her down with a forepaw. Then he climbs on her broad back, balancing on top of her on all fours. She chirps. Her calls apparently arouse Pi's curiosity and he investigates. The medium-sized male rushes at his large rival, who casually walks away, no longer interested in the female and unwilling to assert himself. The male now returns to the female, first mounts her in cursory fashion and after that spends seven minutes happily abandoning himself to her presence. There is nothing austere in his personality as he circles her closely, slides over her, flaccidly drapes himself across her back, paws her affectionately, rolls sideways over her, happily maintaining playful body contact. She crouches tolerantly, while he squabbles with the other medium-sized male. Then the female, apparently tired of these frolics, suddenly hits one male and roars at the other, and when he turns aside, slaps his rump. He retaliates by biting a stick.

Windows in the fog now close, and shaking snow off my coat, I hurry back over the ridge, moving stiffly from the cold yet warmed and elated by the events I have witnessed.

No one knows whom Wei may have tried to court in 1984, and on 9 or 10 December of that year he died. An autopsy revealed that his death could have been due to roundworms that had clogged his pancreatic ducts.

Back in May 1982, after Wei's failure to win Zhen's heart, we transferred our attention to the emerging bamboo shoots. We had established many small bamboo plots the previous year. In these we counted shoots as they appeared, measured them at intervals, and noted how many survived to become stems. Pandas ate many shoots and rodents a few; the larvae of flies and beetles often devoured shoots from the inside, turning them yellow and soft. By these means over a third of all bamboo shoots were destroyed.

One day a bamboo shoot in front of me began to jerk and sway in an animated manner—strange behavior, especially when the shoot, instead of growing taller, shrank until its tip vanished into the ground. Curious, I dug down after it and soon came to a tunnel about six inches in diameter. A bamboo rat burrow. Where soil was deep and well-drained, I often came upon earth mounds that looked as if they had been made by a giant mole.

Such mounds marked the subterranean passages of bamboo rats, two-pound heavy-bodied rodents with large digging claws and soft, gray fur. Once I surprised a bamboo rat in the open. It reared up in a feisty manner as if to box, squinting at me with myopic eyes, its nose pink and flat like a pugilist's, its teeth chattering. I was interested in these bamboo rats because they, like the red and giant pandas, have specialized on a bamboo diet. Sometimes, when an urge to dig overcame me, I excavated a bamboo rat tunnel. The animals pull bamboo into their dark home, eat the stems, and discard branches and leaves by stuffing them into unused passages. To keep its runways clean, a bamboo rat widens the tunnels in a few places and these sites serve as latrines.

Yong Yange, from the staff of Foping Reserve in Shaanxi Province, was in Wuyipeng that spring. Much interested in the work, he accompanied me at times on my rounds, and I was delighted to show him what information we collected. I had been most impressed by his panda research at Foping. There he and his colleagues had patiently approached a female panda again and again until she was used to the presence of people. Without interfering with her natural life, they had observed her behavior even at night. He could tell me, for instance, that in one forty-eight-hour period she drank five times and fed fourteen times in sessions lasting from half an hour to five hours, the kind of detail we could not obtain by radio tracking.

How I wished that our project had a young biologist with Yong Yange's initiative. Hu Jinchu and I would soon depart from Wuyipeng and not return before year's end, and no one dependable would take our places. Wang Menghu had tried for over a year to strengthen the Wuyipeng staff without success.

Because the Cultural Revolution had deprived the country of college generations, well-educated interested young persons were scarce. As noted in a 1987 issue of *Beijing Review:*

> . . . between 1966 and 1969, none of China's colleges or universities enrolled a single student. Many school buildings were turned into factory workshops or barracks . . . numerous lecturers and professors were humiliated and branded "reactionary authorities," and the majority of university faculty members were sent to the countryside, factories or army to do physical labor. . . . Although China's institutions of higher learning began to enroll students in 1970, the national university entrance examinations, however, were cancelled. . . . The university students enrolled throughout this period were called "worker-peasant-soldier" students. . . . Disruption during the "cultural revolution" actually dragged China's higher education back dozens of years and planted a time bomb which exploded in later years as a serious dearth of specialized personnel.

Suitable persons like Yong Yange were either unwilling to come or their departments refused to release them. Only recent university graduates were available. That spring several of them were shipped to Wolong, a posting that was considered the Chinese equivalent of being sent to Siberia.

Kay had left for home in early May, and now I was also preparing to depart. I felt that much had been accomplished during the past nine months, not only in studying the panda but also in establishing cordial working relationships. My presence was accepted, my idiosyncrasies as an individual tolerated, my sincerity in wanting to help the panda recognized and seemingly appreciated. After WWF finally agreed to provide most of the equipment for the research center that the Chinese had requested during the contentious meeting in November 1980, the major disagreement between the two sides had been resolved. Field work in 1983 should proceed without the periodic decline in the level of cooperation that usually signaled a problem between WWF and Beijing. The project still had many difficulties, the principal one being an inadequate Chinese research staff, but these were inherent in the system. As time went on, I came to realize how much genuine dedication, concern, and hard work were given in Beijing and Chengdu to make the project function. All too often, the vigor and effort at the top was negated at the local level. As Millicent Yung, an adviser to WWF who helped the project in many ways, wrote me:

> Each bureaucrat, scientist, technician, administrator, zoo keeper, and student translator has his own allegiance, aspiration, modus operandi. Then the Ministry, the Bureau, the zoo, the Academy, having pledged towards a common goal, has each its own idea how this goal should be reached, and at the same time promoting its own interest. The power of final decision must come from the bureaucrat but his decision can be sabotaged by a student translator, the administrator, the zoo keeper. . . .

The fact that the project was accomplishing its purpose in spite of many problems seemed to show how much China valued the panda.

Next year, I would survey other panda areas, returning to Wolong for only brief periods. Before leaving, I made little pilgrimages. I ascended the mountainside to the tent at 4X where Kay and I had spent many nights monitoring pandas; others would now do these routine tasks. Such locations remain fixed in the mind, and I would always remember the curves of the trail, the outline of the hills. The fact that I looked at the spot with nostalgia was a reminder that one tends to complain of discomfort but relish the memory of hardship. I also clambered to the rocky top of Bai Ai to see the special kind of crimson rhododendron that bloomed there at this season. To the east, a waterfall of cloud cascaded over the ridge into the Choushuigou, yet I was in sun, the valley divided between gentle repose and roiling violence.

On my way to Chengdu on 6 June, I stopped at Hetauping to photograph for WWF the progress being made in construction of the research center. A week later, I wrote WWF a letter that concluded: "I am sending under separate cover some photos of the research center. The walls are up on the research building and the foundations of the bridges are under construction."

On 13 August, WWF sent a letter, signed by its top leadership, to Zhang Shuzhong of the Environmental Protection Office, and it read in part:

> Due to the uncertainty surrounding construction of the research center, our National Organisations have not been able to seek contributions specially earmarked for this purpose and, as you know, it is our policy not to use general conservation funds for construction.

WWF then offered to start a special campaign to raise funds "provided we can be assured that construction has been started." And it went on to note:

> We must, however, make it clear that, while we can assure funding for a continuing Research and Conservation Programme, we cannot guarantee that we shall be able to raise sufficient funds to cover all the capital construction and equipment costs which had originally been agreed upon. . . . In conclusion, we should like to assure you of our desire to continue close and fruitful cooperation.

The Chinese and I were nonplussed. The message of the letter seemed clear. WWF wanted to break its agreements, the protocol and action plan signed in 1980, and renege on the remaining six hundred thousand dollars for construction of the research center, and by implication, on the equipment as well.

Any trust that had developed between WWF and China in the past year was again shattered, the future of the field research once again in doubt.

Death in the Choushuigou

April 1983

I had hoped to continue field work after Spring Festival in February, starting with surveys of several panda habitats. However, when WWF wrote China that it could not meet its obligations to the research center, my 1983 program was promptly canceled. A December 1982 meeting between WWF and the Chinese failed to resolve the financial crisis. Later, however, WWF promised to pay an installment of four hundred thousand dollars in April. A new agreement would be signed at that time, and I was notified that field work could then continue. Eager for news of Zhen and the other pandas, I pleaded for permission to return to Wolong before the new agreement was signed. The Chinese generously consented to my return in late March. In Beijing I met with Wang Menghu and others. Showing restrained anger, some genuine and some simulated to emphasize points, they expressed at length their disillusionment with WWF. Toward the end of the meeting, Wang Menghu said, "I have bad news for you. Don't get tense when you hear it. Remain calm."

With such a preamble, I expected to find myself on the next plane home. But the news was even worse.

"Han-Han was killed on 24 January. A report was submitted to the State Council. Premier Zhao Ziyang says the case should be investigated."

Beautiful Han-Han was dead, strangled in a poacher's snare one year after we collared her. Hu Jinchu met me in Chengdu and told me the details of the incident.

The Wuyipeng staff began to monitor twenty-four hour activity on 23 January. When at 11:15 the following morning Han's radio signaled that she was inactive, the change seemed no more noteworthy than the beginning of any rest period. However, when the signal remained inactive the rest of that day and the following night as well, the staff assumed that she had removed her collar, and several persons went to retrieve it. After searching for two days they found the collar hidden beneath a rock. Nearby were Han's viscera covered with snow and leaves. The staff had inadvertently monitored Han's dying moments as, caught with the wire snare around her neck, she slowly strangled. On finding her body, the poacher skinned and butchered it and carried the meat and hide to his home in the valley. The government responded swiftly to the report of Han's death. Several men from Chengdu's Public Security Bureau arrived with two German shepherd dogs. The dogs tracked the poacher downhill to his home, where he was arrested and Han's hide confiscated.

Hu Jinchu also noted that, judging by Han's activity cycle the previous September, she probably had a baby, though no one had checked the den area then or subsequently. At the time of her death, her baby would have been too young to survive on its own.

Han died in the same area where I had found snares during the previous two winters. My repeated entreaties to control hunting and illegal woodcutting within the reserve had not resulted in action. When I first heard of Han's death, my thoughts immediately went to Ning-Ning. We had last heard her radio signal on 30 March 1982, not far above the highest fields. Transmission ceased abruptly from one day to the next, and we never picked up her signal again. The telemetry equipment, manufactured by Telonics of Arizona, was excellent, and a radio was unlikely to malfunction that suddenly and completely after only three months on an animal. I felt certain that Ning had also died in a snare. The poacher had perhaps unknowingly cut the antenna inside the collar when he removed the radio to hide it, greatly reducing transmission distance.

Shortly after Han's death, on 8 April, I was in the area where she had lived to survey the damage done by woodcutters. All winter long, I had been told, the sound of axes could be heard. An ancient Qiang song begins, "The greatest are Heaven and Earth, next to them is the Sacred Forest. . . ." But judging by the many raw stumps I found, these words were now ignored. As I followed an animal path, I spotted the silver of a panda's back, the animal crouched asleep by a clump of bamboo. When I moved quietly to one side to get a better view, the realization that the panda was unusually still suddenly resounded in my mind. It was dead, its neck in a snare. I stood by the body, numb with the shock of the living in the presence of sudden death. The motionless form of the panda expressed such a suffering innocence that its power was almost religious in intensity. For the first time I directly confronted the tragedy of a panda's death. Despite my grief at Han's death, it had been mainly an intellectual pain; I had not actually witnessed her vulnerability. Kneeling by this panda, its death so empty of purpose, I was filled with rage—at the unintentional killer, and at the officials who talked and talked while ever more pandas slipped silently into darkness. "Damn. Damn. Damn them all."

My indignation heightened as I examined the panda, a female, more closely. The snare was tied to a sapling so small she could have severed it with one bite. Instead she had gone round and round a nearby bamboo clump until she choked, and in the throes of death bit off the tip of her tongue.

I reached out and stroked her back. Her body was cold.

The next morning about a dozen of us, including Hu Jinchu and the Wolong police, returned to the body where we stood in a semicircle before her, not speaking, looking down at her still form as if in prayer. It was a silence far louder than the usual chatter. Finally, as if in slow motion, we took pliers to unwind the snare from her neck and turned her over. There was milk in her swollen nipples. She had died at the height of life's success; somewhere in the forest was an orphaned young about seven months old, still in need of milk, but with luck it might survive on bamboo leaves. We tied her to a pole, and two men carried her down the mountain, one of her arms bobbing in rhythm to the steps. The rest of us followed in a sepulchral procession until we reached a hamlet of six houses by the road. Leng Zhizhong, the man who had killed Han, lived here. Villagers lined the path to watch us pass and load the body on the truck waiting there. A team from the Public Security Bureau had arrived with tracking dogs, but no one was ever arrested.

In one year, two and probably three female pandas in our small research population were known to have been killed by poachers. I felt guilty that such crimes could still exist in spite of our efforts to help the panda. My hopes for the study were now phantoms; idealism had given way to savage pessimism. But one cannot give in to despair: the animals need someone to fight for their right to exist. If pandas elsewhere had the same problems as at Wolong, then the species was withering away, an unrepeatable evolutionary act destroyed in the silent forest.

Pandas had been heavily hunted in China before Liberation in 1949. Donald Carter, for instance, described a spear trap he saw in the 1930s:

> A spear with an iron head and a wooden shaft is placed horizontally, at the required height, between two upright sticks. To the end of the shaft is attached a sap-

ling which is pulled back and caught with a trigger. This sapling forms the spring. A cord runs from the trigger across the trail upon which the panda is supposed to travel. When the cord is tampered with the spear is driven forward with tremendous speed, guided by the upright stakes. The spear is so set as to strike near the animal's heart.

And in the 1940s, William Gruenerwald saw many panda skins for sale, as he related in a letter:

> During the latter part of WW II, I was involved in mineral surveys in western China and Tibet. One of the outfitting sites was the town of Kangting—then called Tatsienlu—located along the 'marches' of Tibet. It served as a transfer point for the tea trade between the coolie carriers of Szechwan and the Tibetan yak caravans. The main street was a marketplace for traders with all the wares of China and Tibet for sale and barter. I was amazed at the collection of animal skins and pelts for sale: Siberian tiger, golden monkeys, stacks of snow leopard, antelope and panda skins. One merchant had a stack of 5 panda skins to choose from!

I certainly had not expected to lose animals to poachers in a reserve, much less in the area designated for international cooperation in panda conservation.

I had a meeting with Lai Binghui, a Party Secretary who had become Wolong's leader after Zhao Changgui left. Normally jovial, Lai Binghui was subdued as he said:

"We are not strong enough to restrain minorities; we have been too lenient. If we are strict we are accused of racial discrimination. And we still have internal policy problems."

Replying that I understood the situation, that the United States had similar situations with Indians and Eskimos, I then emphasized that urgent action was needed to protect Wolong's pandas. Losses were too great, and recovery would be very slow. I recited figures to show the exceptionally low reproductive rate of the panda. A female can at best raise one young every two years. Since young frequently die, a female is lucky if she raises one young every three years. Although a panda may live twenty-five to thirty years in captivity, few probably reach even twenty years in the wild. Therefore, if a female has her first young at the age of six years, she would by the age of eighteen years have been able to raise only about five young. If losses in the Choushuigou are an indication, Wolong's panda population is declining rapidly.

Once again I repeated my litany about the need to send police patrols into the forest to control poaching. Beguiled by an easy answer, Lai Binghui deflected my comments and stated, "We will provide the local people with necessities so that they will not cut trees illegally or hunt. We will make them rich, at great cost to the government."

Han's death was in vain.

To emphasize the importance of Han's death and to serve as a lesson to the local villagers, the People's High Court of Sichuan moved to Wolong for the poacher's trial on 22 April. I was invited to attend.

In China, most crimes are handled by mediation committees, and relatively few cases reach court. Those that do enter the court system almost invariably result in a guilty verdict.

Several hundred villagers and other Wolong residents crowded into the meeting hall at headquarters. Everyone not only knew that the defendant, Leng Zhizhong, would go to jail, but also for approximately how long. Yet there was an expectant hush as the court came to order. Village men in goatskin vests puffed intently on their small brass pipes, and even the babies, swathed tightly and strapped to their mother's backs, were silent.

The judge, with two assistant judges and a secretary, as well as the prosecutor and defense attorney, sat down behind tables on the raised stage of the meeting hall. The secretary stated the charges, amd the judge, dressed in blue cotton, said, "Bring in the accused."

Two uniformed policemen, revolvers strapped to their waists, marched the culprit, Leng Zhizhong, twenty-six-years old, before the judge. Of medium height, with a broad pleasant face and a thin mustache, Leng Zhizhong wore blue cotton like the judge, but his collar was adorned with artificial fur. He shifted nervously from foot to foot.

"The accused understands that he does not have to speak," the judge stated. "But if he wishes, he may speak on his own behalf."

The indictment was then read.

"The accused was arrested on 21 February 1983. Based on material submitted there was a pre-trial hearing on the first of March. The accused is educated; he knows that rare animals are protected. He killed a radio collared panda with a snare. He met three people on the way home who knew the consequences. Yet he did not report the incident but went back with a knife, cut the collar, skinned the panda, and cooked the meat. He defied the law. His crime is that he destroyed a rare animal."

The judge then questioned Leng Zhizhong.

JUDGE: "What is your occupation?"
LENG: "Farming in Wolong."
JUDGE: "By what means did you kill the panda?"

LENG: "Wire snare."
JUDGE: "Where did you get the wire?"
LENG: "Cut it off a bridge."
JUDGE: "When did you set the snare?"
LENG: "January 3, 1983."
JUDGE: "How many snares did you set?"
LENG: "Sixty to seventy."
JUDGE: "Did you know that you set the snares in a forbidden area?"
LENG: "Yes."
JUDGE: "Why did you set snares?"
LENG: "In December I went to cut firewood in the mountains and saw many snares and therefore thought it was all right to kill animals there. I wanted to get animals to sell to settle some accounts. I made-believe I went to cut firewood. The snares were hidden under my coat."
JUDGE: "What did you want to catch?"
LENG: "Musk deer. And also wild pig."
JUDGE: "Tell me of the discovery of the panda."
LENG: "I saw footprints in the snow. At first I thought they were those of persons. Then I found the strangled panda. I know it is wrong to kill a panda and wanted to conceal it. With a knife I cut off the feet; then, I opened the body and took out the insides. There was a thicket and I hid them there. I took away the meat and skin."

The knife was presented in evidence, a butcher knife with a ten-inch blade and a sheath of untanned hide.

JUDGE: "Did you see anything special on the body?"
LENG: "A radio collar."
JUDGE: "What did you do with it?"
LENG: "I cut the strap and hid the collar beneath a rock."

The radio collar with its cut strap is presented as evidence. A witness who met Leng Zhizhong on that fatal day was called. Round and red-faced, he laughed harshly as he took the stand.

JUDGE: "What did the accused say when he met you?"
WITNESS: "I went to cut bamboo. When I asked if he had caught anything, he said no.

After that, Han's hide was presented as evidence, feet and head cut off. The judge turned back to Leng Zhizhong.

JUDGE: "What did you do with the meat?"
LENG: "I carried it home and my wife cooked it with turnips. We ate some. It did not taste good. So we fed it to the pigs. And I took some to my sister."

A statement was read to the court describing how police tracked the accused from the site of the crime to his home on 5 February.

JUDGE: "What do you want to say on your own behalf?"

LENG: "I committed this crime. I am poor and did it to help my family. I caused a great loss to the people. I understand this now and will not do it again. Selfishness is the cause. As a result of being in prison, I will be helped to walk a new road."

Having asked these and other questions, the judge noted that the investigation was completed. The case would now be turned over to the prosecutor and the defense.

The prosecutor rose and presented a narrative review of the case. He concluded as follows.

PROSECUTOR: "The criminal has confessed to everything. The crime has caused social harm. The panda is rare, beautiful, beloved by all people, and well known throughout the world. It is a valuable animal in scientific research; its death will affect the study. It is forbidden to kill in protected areas. Rare animals must find protection in natural reserves. The accused knew that he was not allowed to hunt. If a snare can catch a boar, it can catch a panda. The crime is indirectly intentional. This is a lesson that must be learned not just by the accused but by others here as well. The leaders of Wolong were not involved. Still, they have a lesson to learn: there are gaps in their work. Even though the commune members are poor they must obey national laws."

The defense attorney first made an eloquent plea for conservation, noting that, in many countries, environmental protection has made great strides in recent years. China, too, has laws to protect rare animals, and it has one hundred and four natural reserves including Wolong, which has been designated as a World Biosphere Reserve. The death of a panda, he continued, was a great loss to Wolong, to China, and to people all over the world, but there are mitigating circumstances that should be given consideration by the judges.

DEFENSE COUNSEL: "The newspapers say that the evidence is strong against the accused. It is my duty to give evidence that his crime is not so serious, that he is innocent. That his snare strangled the panda is not in doubt. But the motives of the crime must be considered further. His crime is not isolated: he saw many other snares. His intention was only to catch musk deer like the others. He should be punished. But remember his family is poor and he committed a crime accidentally. The accused has only a primary education; he does not understand ecology. The purpose of our laws is to educate, not punish. There are several thousand snares in the reserve. The danger to the panda persists. This is a thought-provoking problem. Our propaganda obviously does not keep pace with the actual conditions. The peasants here are poor and we must improve their conditions. However, I hope they realize that the country has problems too. We must not only prevent the killing of wildlife but also educate the people and raise living conditions. This will change the irresponsible into the responsible."

There was a ten-minute recess after the defense attorney concluded his statement. Then, the crowd having convened again, the judge announced his verdict.

"The sentence is two years in prison. The accused can appeal his sentence in ten days."

Although everyone already knew the verdict, the audience still gasped. The two-hour trial was over.

I thought the proceedings had been impressive. But I also thought that the wrong person had been sent to prison. It was not Leng Zhizhong who should be in jail, but those local officials whose negligence and stout indifference had made them accessories not only to Han's death but also to the strangled female I had found. Leng Zhizhong would "be helped to walk a new road," but the officials would merely walk back to their offices and their unchanged routines.

Travels in Panda Country

April to June 1983

Geographically, Sichuan is a world of its own; mountains ring the central basin and in the past only treacherous paths crossed this barrier to the rest of China north, south, and east. The Yangtze River, however, has always offered a tenuous link to the outside, though its gorges and rapids prevent easy passage. At the basin's western rim begins a chaotic maze of mountains, a tossed sea of sharp ridges, snow peaks, and gloomy valleys that extend to the high plateau of Tibet. Since the first Chinese settlements in the basin over two thousand years ago, the population had increased to fifty million in the 1930s and to one hundred and ten million today, a growth made possible by good soil and a mild climate that permit up to three harvests a year. Despite its isolation, much of Sichuan through the ages has remained culturally an integral part of China. Not so the mountains in the west. Once called the Sino-Tibetan borderland, the region lacked a clearly defined boundary between China and Tibet until the 1950s, when China consolidated its power.

In the spring of 1983, I was allowed for the first time to travel widely in this borderland to survey pandas (see map, p. 232). The excitement of visiting a new area is often based on its history. Père David had obtained the first giant panda skin there and other well-known Westerners—explorers, museum collectors, travelers—had crisscrossed the mountains or followed the major trade route to Tibet that went from Chengdu to Kanding (Tatsienlu) and Batang on to Lhasa. In the southern part of the panda's range live the Lolos, or Yi as they are called today, a people traditionally hostile to all outsiders. Many tribes of non-Chinese inhabit this region, and in times past they were divided into feudal states governed by hereditary rulers. Some, such as the Yi, were independent, but most came under the strong influence of or even paid tribute to China after its armies crushed a rebellious confederacy in 1775 and broke the power of the tribes. Throughout this frontier region the Chinese had trade centers and built military posts in strategic places, controlling especially the major trade route to Tibet. Now as then, Chinese dominate in the population centers. Tribal peoples inhabit the hills, where in small villages and isolated farmsteads they struggle to cultivate crops, raise livestock, and use the forest for cutting timber, collecting medicinal roots, and hunting wildlife. Although many Westerners visited the area until the 1940s, and there were even Christian missions in a number of towns, the region had retreated into anonymity. I was intrigued to discover what changes the decades had brought.

24 April. I am exhilarated. On this day for the first time in my two and a half years with the panda project, I am leaving the confines of Wolong to visit other panda habitats. Our goal today is Baoxing, a county adjoining the Wolong area. We head southwest of Chengdu, Hu Jinchu, Xiao Qiu, the forest official Cui Yangtao, and I, first across the plains of the Red Basin and then through hills terraced with cultivation to the rounded summits. The plots of wheat, rape, and vegetables are immaculate and attest to the efficiency of the farming methods. That China, historically plagued by famine, can feed its billion people on just one-third of an acre of arable land per person is an achievement of tremendous magnitude. Much of China's rural life takes place in the "free markets" by the roadside, where people sell for profit whatever has not been contracted to the state. Stalls and carts there overflow with vegetables, fruits, eggs, chickens, and pigs, as well as homemade goods, from brooms and baskets to vinyl-covered sofas. Vendors sell candy and cigarettes, bars of soap and sunflower seeds, and there are freelance barbers, cobblers, and bicycle repairmen. A new economic vision spurs the populace on, the Eight New Things—wristwatch, color TV, camera, washing machine, cassette player, and other items—part of everyone's new dream.

Beyond the town of Ya'an, a dirt road traces a riverbank westward toward a wall of mountains where a dusky gorge opens like a gateway into a broad valley. Beyond, a pass offers entry into the mountainous interior of the Sino-Tibetan borderland. We drive down a valley hemmed in by peaks, their summits in cloud. Fields have been slashed unterraced into the steep hillsides, the furrows vertical to permit heavy rain, and with it the soil, to run off; all along the road, farmers are collecting this soil in baskets and laboriously carrying it back up to their fields.

Coming around a bend in the road, we see Baoxing town ahead, crowded between mountain and river. It is a typical county town with a cluster of undistinguished cement buildings, each several stories high, comprising government offices, a department store, hostel, and movie theater; small businesses and homes straggle toward the countryside. Baoxing, once known as Muping or Moupin, had been one of many feudal principalities between Chengdu and the Tibetan frontier. About five hundred years ago, the Chinese tried to drive out the original inhabitants, but when they were unsuccessful, they persuaded some of the eastern Tibetan tribes to invade, and these tribes thereafter became the rulers. By this century, their power had essentially vanished, as noted by the Roosevelt brothers, who had hunted pandas here in 1929.

> From time immemorial it [Muping] had owed some shadowy allegiance to whomever might be its nearest powerful neighbor. Close as it is to Chengtu, which has been a great Chinese city since long before the days of the three kingdoms, it is remarkable that it retained its individuality so long. Its annals are a continuous record of struggles against outside aggression. The curtain dropped on the drama only last summer. The ruling family was Tibetan. The last prince died some years ago, leaving a wife and one daughter. A match was arranged with a prince of the Chalu family of Tatsienlu, who came to Muping. About a year ago he was killed, either by his own vassals or the Chinese,—reports differ. To-day there is a Chinese magistrate in Muping. . . . There were the remains of past glories in the shape of a tumbled stone lamassary,—now the soldiers' barracks,—an old fort crowned a near-by hill, and the ruined footings of what had been the king's palace, with two big stone lions guarding the entrance. . . .

By the side of the road on the outskirts of town, I find a stele with its ancient inscription obscured by the passage of time and the arrival of a new era. Incised over the old characters is a current message: "Only communism can make the people great."

To any Westerner concerned with pandas, Baoxing is a place of pilgrimage, for it was from here that Père Armand David ventured into the mountains and discovered the giant panda. A Basque from the Pyrenees, Armand David was born in 1826, and at the age of twenty-two he entered the Congregation of Lazarists. He was ordained in 1851, and eleven years later he was sent to the mission of the Lazarists in Beijing, where he applied himself so zealously to the study of natural history that his discoveries include some of Asia's most remarkable animals and plants. By 1865 he had already seen an "interesting ruminant," a strange-looking deer, in the emperor's imperial hunting ground near Beijing. Called *sibuxiang*, the "four unlikes," because it seemed to have the tail of a donkey, antlers of a deer, neck of a camel, and hooves of an ox, the deer was extinct in the wild and this herd was the last one in existence. Père David bought the remains of two deer, later named in his honor *Elaphurus davidianus,* Père David's deer. His second collecting journey, which lasted from 1868 to 1870, took him to Baoxing, and here his discoveries brought him not just scientific recognition but also popular fame.

In Père David's time the trip from Chengdu to Baoxing required six days, the final day a trek across a high, thickly wooded crest. Descending toward Baoxing, Père David saw "many ancient trunks of a very big *sha-mu* (pine) rotting on the ground, having been felled by order of the *T'u-ssu,* or prince of Muping in order to erect a barrier impassable to Chinese troops." At this time, there was one of the Missions Étrangères near Baoxing, run by a French missionary, a Monsieur Dugrité, who taught about fifty Chinese students "Latin, philosophy, theology, history, and so on," and Père David at first stayed with him.

From Père David's journal:

> March 1 1869 Monday, first day in Muping. Very fine weather. As I have already noted, the college at Muping was founded fifty or sixty years ago when persecutions raging in China obliged the missionaries to find a safe refuge in the estates of the prince of the Man-tzu. At the time, these valleys were still entirely wooded and the only natives in the country were called barbarians. But, before long, Christians and other Chinese followed the missionaries, and obtained permission from the local sovereign to cultivate the land under certain conditions and levies. Gradually the valley took on a Chinese appearance, Chinese agriculture was introduced and the missionaries brought in potatoes and cabbage from Europe, two plants which today constitute the principal food of these mountaineers.
>
> The culture of medicinal plants, the hunting of the musk deer, and the making of potash by burning plants, along with the culture of corn and plants already mentioned, provide the living, such as it is, of the inhabitants of these inaccessible mountains.
>
> The cutting and transportation of wood sawed into wide planks is also a resource of the country.
>
> As soon as I am settled in the comfortable little room which M. Dugrité put at my disposition I hasten to see the surroundings of my new home, where I expect to stay a whole season. . . . I spend every free moment of the first day hunting for birds. . . .

I can understand Père David's eagerness to explore the forest, to satisfy his curiosity. There is a photograph of him in a Chinese cloak and fur hat. He wears a black goatee, his eyes are piercing, his features are alive and determined, and a touch of humor plays around his mouth; his face conveys resourcefulness and quiet passion. Although he was a collector of wildlife, he also wrote, "Man is the king of nature, but he has not the unpleasant right to be the butcher except when this is absolutely necessary." His insight into nature, in effect his personal creed, as quoted at the beginning of this book, could have been written by one of today's environmentalists. What a pleasure it would have been to explore the mountains with this man.

25 April. Crags and a patchwork of fields and brush on steeply tilted slopes flank the road along the Dong River on the way to Yaoji. We halt at a side valley, the Dengshi Gorge, with a "Catholic temple," as Cui Yangtao calls it, where Père David had stayed. I have been given special permission to visit the site, Cui tells me, the first foreigner allowed to do so since the 1940s. Below our foot trail, the white foam of a stream crashes against boulders. A cinquefoil and iris are in bloom. Soon the slopes retreat and the valley veers toward the north, all but the summits cultivated with rape, potatoes, and maize. Scattered huts with wood-shingled roofs perch on terraces, and chestnut and walnut trees border our path as we angle upward. After a good hour's walk, on turning a corner in the trail, we come upon the temple, an imposing wooden building with sweeping Chinese roof, large yet not intrusive as it nestles against the slope with a view of snow peaks a day's journey up-valley. Here at nearly six thousand feet, the sun is warm, the breeze crisp. Here the panda's history, past and present, comes together.

Massive double doors seal the temple against casual intrusion. A large sign above the door proclaims "To serve the people," a slogan appropriate to all ages. A caretaker swings open the gates and we enter a courtyard. Painted characters on one balcony exhort, "We must increase our output to three thousand tons." For several years the temple had been used as a base by workers of a nearby asbestos mine, but at present the rooms stand empty and only the caretaker lives here. He is unhappy, for when the winds howl down from the summits, ghosts in the building creak and moan. Many beams in the temple are charred, a reminder of decades past when locals burned part of the building and missionaries rebuilt it. The chapel is large and the ceiling high, its ribbed vaulting of inlaid wood. Light filters softly through the fine latticework of the lancet-shaped windows; one window is large and round, with a latticed design that vaguely resembles a Buddhist mandala with a central cross. Once there were many statues, the caretaker says, but they were smashed during the Cultural Revolution. Logs and mining equipment now occupy the chapel.

Outdoors, I walk through a garden overgrown with grass and vines and sit on the remains of a rock wall, the memory of Père David around me.

This was his base, he tramped these hills more than a hundred years ago; from here he made several of his greatest discoveries—the dove tree, the Chinese monal pheasant, the golden monkey, and, as related in his journal, the giant panda.

March 11 Upon returning from an excursion we are invited to rest at the home of a certain Li, the principal landowner in the valley, who serves us tea and sweets. At this pagan's I see a fine skin of the famous white and black bear, which appears to be fairly large. It is a remarkable species and I am delighted when I hear my hunters say that I shall certainly obtain the animal within a short time. They tell me they will go out tomorrow to kill this animal, which will provide an interesting novelty for science.

Père David refers to "*du fameux ours blanc et noir,*" usually translated as "famous white and black bear," though there is no evidence that he knew much about the animal. "Fameux" can also mean "first-rate" or "capital," perhaps a more appropriate translation.

On 23 March, Père David received a young panda, just killed by hunters, and on the first of April an adult. He felt that the panda "must be a new species of *Ursus* [bear], very remarkable not only because of its color, but also for its paws, which are hairy underneath, and for other characters."

Since 1949 the Baoxing region has provided more pandas to zoos—including Hsing-Hsing in the National Zoo, Washington, D.C.—than any other, a total of one hundred and one pandas by the early 1980s.

At Yaoji, mainly a Tibetan community, the terrain is gentler and the hillsides are more arid than around Baoxing. The town consists of a short main street lined with wooden buildings, most of them shops, various government barracks, and, on a knoll, the ruins of a temple. Houses are Tibetan in style, two stories, the lower of stone and the upper of wood, often with a balcony, the architecture much like that found in the Alps. Women here wear the usual Tibetan ankle-length cloak, or *chuba;* their headdress is distinctive, a stiff black cloth in multiple folds, decorated with amber, coraline, turquoise, and silver, and held in place by a braid of hair. To explain the purpose of our visit, we meet with the local leaders, and as we talk the hirsute heads of Marx, Engels, Lenin, and Stalin as well as the clean-shaven Mao gaze down upon us from posters on the walls.

26 April. The morning after our arrival we ascend a valley, the Nibaugou, to one of the logging camps, where pandas have frequently been seen and where they even come to the huts in winter to scavenge among the refuse. With us is Gao Huahang, a young forester from the nearby Fengtongzai Reserve whom I remember for his congeniality and wide smile when he assisted at Wuyipeng. Last year he had observed a panda female in her den, and he wants to show us the site. After threading our way upward through a chaos of logs, severed branches, and mangled undergrowth, we reach the crest of a spur along which a remnant strip of forest persists. There, just twenty feet from a busy foot trail, is the den, a fir with a hollow base and an entrance fifteen inches in diameter, barely large enough for a panda to squeeze inside. The female had given birth there on 24 September and had remained at the den for a month, even though loggers passed daily to cut timber on the opposite slope. Sometimes they fed her meat and bones. Gao Huahang showed me a photograph of himself taking notes in front of the den, the female inside peering up at him. Desperate for a den site, the panda had adapted to people and the sound of the forest crashing down all around her. On the way back, we inquire of workmen if they have found recent evidence of pandas. Yes, one was heard calling yesterday above us on the slope. Followed by Xiao Qiu̧ and Gao Huahang, I clamber up. Ahead, along the ridge, I see a panda lumbering away past an aerial cable driven by loud motors that is being used to haul timber down the mountain. The valley is filled with the din of chopping and shouting.

Later, near the lumber camp, we hear the squeal of a panda, and fifteen minutes later another. But the animal detects us, and only swaying bamboo reveals the line of its retreat. All this human activity has obviously not deterred pandas from remaining in the area.

I note with interest that two kinds of bamboo predominate here. One, *Sinarundinaria chungii,* tall and graceful like umbrella bamboo, extends up to an elevation of nine thousand feet, and the other, arrow bamboo, thrives at higher elevations. Many stalks of arrow bamboo are in bloom, the inflorescence at the tip of each branch resembling that of rice, and it is purplish brown in color. These flowers, so innocuous in appearance, herald potential disaster, a threat to the panda's precarious existence. Pandas appear to have little to do but eat and nap and court. Yet nature seems to have extracted a terrible penance for such a sybaritic existence, for at long intervals the panda may starve when the bamboo blooms and dies.

Bamboos belong to the grass family, though they are unusual in having woody stems and branches. Unlike typical grasses, which flower and produce seeds annually, many bamboos flower only at long intervals, reproducing at other times by sending out shoots from rootlike rhizomes. Periods between flowering vary from species to species, from as little as fifteen years to as long as one hundred and twenty years. After flowering and seeding, the bamboo plant dies. Sprouting seeds ultimately form new clumps of bamboo. Several new stems are added to a clump each year and old ones die, and no stem lives more than ten to fifteen years. On occasion a solitary clump of bamboo may bloom, but usually almost all members of the same species in a mountain range or region burst into flower at one time, and the stems then die in synchrony. Each bamboo cell must contain a genetic clock that signals the time when a plant will switch from one reproductive method to another. Since intervals between flowering are not wholly precise, it has been suggested that an environmental factor—perhaps sunspots, drought, or earthquake—triggers the change. Yet these factors are not that predictable, and not all bamboo species die in synchrony. Furthermore, when a bamboo clump is transplanted to England, for example, it may flower at the same time as its kin in China.

This spectacular mass flowering has long elicited comment in Chinese literature. A Qin Dynasty (221–206 B.C.) book on bamboo noted that "after 60 years it bears seeds and dies." And there is a late seventeenth century prescription in an agricultural manual for preventing bamboo from dying:

> Whenever bamboo flowers, then all the bamboos in the plantation will die. The way to deal with this is as follows: Take the largest stem and cut it. Leave three feet in the ground. Bore through all the sections and fill the hollow with manure. This will immediately stop the flowering.

While this technique might be effective in a garden plot, the method could be cumbersome on a mountainside of reedy arrow bamboo, where a hundred million stems can crowd a square mile.

During the mid-1970s, bamboo died widely in the Min Mountains of northern Sichuan and southern Gansu; about two thousand square miles were affected to varying degrees. Such widespread flowering there had last been observed in the mid-1880s by the Russian M. Berezovski, among others. The flowering in the 1970s might have aroused excitement only among China's bamboo specialists, were it not for the fact that the panda had become a national treasure, and the animals were starving to death. Little could be done to help them. "After surveys and investigations, a total of 138 panda bodies were found," wrote Wang Menghu, "What a heartbreaking loss."

Was the flowering of arrow bamboo confined to this one hillside in Baoxing County? I doubted it. Two years earlier, in May 1981, I had found small patches of arrow bamboo in flower above Wuyipeng. I had thought then that the species would soon mass-flower, seed, and die. Besides, arrow bamboo was due to seed again for, according to local people, it had previously flowered and died in 1893 and 1935. Memory of these dates was precise because the former date was associated with a peasant rebellion, and the latter with the Long March, when Mao Zedong and his followers had passed west of Wolong on their way to Shaanxi. Certainly the pandas now faced a crisis in Wolong and other parts of the Qionglai Mountains.

3 May. The waters of the Min River are silt-brown and in spring flood. On our left, from an area bristling with peaks, the Tsaopo valley joins the Min. Several of the early panda expeditions had shot their animals in the Tsaopo—the Dolan expedition of 1931, the Sage expedition of 1934—and both Ruth Harkness and Floyd Tangier Smith had obtained animals for zoos from there. Farther north along the road, where the Min valley widens, is Wenchuan, once the capital of the principality of Wassu, center of the local Qiang tribe. Now, however, it is a modern county town, and even "the ancient and broken-down city wall," as William Sheldon described it in the 1930s, is gone.

For over two thousand years the valley of the Min has served as a major route for traders and invaders, and only the ruins of forts and watch towers on ridge tops remain a medieval memory. The hillsides must have been forested then, different from the desolate gray and dry slopes with scant tufts of vegetation that we drive past. Spring seems to have bypassed this arid land. Although it is May, grasses and shrubs retain the dun color of winter, but near a hamlet further on, apple trees are in blossom. Dusty and rough, the road permits only weary progress; a fist-sized stone hurtles from a cliff, smashes onto the car hood, and cracks the window. At the Songpan Forest Bureau, a huge timber operation high in a side valley among conifers and birch, the leaders greet us with a lavish banquet, and afterward we watch television, a ping-pong match between Sweden and China, a reminder that in today's world remoteness is often illusory.

4 May. Further up the Min valley, at an elevation of nine thousand feet, the river becomes small and clear, and the air is crisp beneath a sky that has the crystal clarity of Tibet. On a plain of spring-green fields lies the town of Songpan. Beyond a crowded market rears the old city wall, twenty to thirty feet thick and about as high. Yaks and shaggy ponies laden with goods jostle down the main street. Women wear wine-red *chubas,* held together by belts studded with silver. Their head adornments are distinctive, as they are in each Tibetan tribe, red turbans decorated with a corona of amber-colored beads the size of walnuts.

Songpan was established in 1775 by Emperor Qianglong as a military post to keep the local Tibetans, the Sifan, under nominal control, and soon it became an important trade center. Ernest Wilson, who visited here several times during the early 1900s, wrote:

> Did the Fates ordain that I should live in Western China I would ask for nothing better than to be domiciled in Sungpan. Though the altitude is considerable the climate is perfect. . . . Excellent beef, mutton, milk and butter are always obtainable at very cheap rates. The wheaten flour makes very fair bread, and in season there is a variety of game. Good vegetables are produced, such as Irish potatoes, peas, cabbages, turnips, and carrots, and such fruit as peaches, pears, plums, apricots, apples, and Wild Raspberries (*Rubus xanthocarpus*). Nowhere else in interior China can an Occidental fare better than at Sungpan Ting. With good riding and shooting, an interesting, bizarre people to study, to say nothing of the flora, this town possesses attractions in advance of all other towns in Western China.

North of Songpan, the hills become low and rolling, their contours broken by stands of spruce and cedar, and at this season they are still partially snow-covered. The river becomes a mere creek meandering over gravel bars bordered by willows. From a pass at nearly eleven thousand feet, the crest of the Min Mountains, the road descends in tight curves, and the rugged limestone peaks with forests of spruce along their flanks deprive us of horizon. At ninety-seven hundred feet, I see the first bamboo, graceful green clumps of *Fargesia nitida*. Further down the slope this bamboo is in bloom, all its stems brown and leafless, each branchlet crowded with flowers. The valley joins a larger one, the Tazang, and we turn eastward, passing through cultivated lands until we reach the entrance to the Jiuzhaigou Reserve, our goal.

A clear stream tumbles down the valley flanked by pine forest, its descent interrupted by several natural dams, the result of ancient earthquakes. Behind each dam a lake has formed, each an intense turquoise color and so clear that one can see submerged logs resembling shadows of crocodiles near the bottom fifteen feet or more below. The lakes may spill over the dams in cascades, in splashing leaps, or the water may simply scurry among islands of willow. Ahead, limestone peaks thrust into the sky like the teeth of an inverted saw. The scene has grandeur, yet its harmonious beauty gives it intimacy. At a fork in the valley the reserve headquarters has been built, as is all too often the case in Chinese reserves, with an unerring ability to select a sublime site and deface it with brick-and-cement structures derelict in appearance.

Eager to explore, I hike along the shore of a nearby lake. My route is littered with broken bottles and fruit jars—Chinese tourists have a penchant for smashing glass—plastic bags, empty film containers, and other refuse. Veering up the side of a mountain, I come to the base of a cliff where a large stand of bamboo is in flower; in fact all bamboo below an elevation of eighty-five hundred feet is blooming in synchrony. Above this elevation, the bamboo died in 1976, according to the reserve staff. As only one bamboo species occurs in the Jiuzhaigou, its different flowering cycles at different elevations helps explain how pandas have been able to survive the periodic bamboo die-offs: the animals can simply shift up or down the slope.

5 May. In the morning we drive up one of the valleys, the Chawa. A large Tibetan village crowds against one hillside, prayer flags fluttering from tall poles. Some eight hundred Tibetans live, hunt, and cultivate crops in the reserve. The road climbs steadily up through a canyon and then past slopes denuded of spruce and fir; bamboo, much damaged by the logging, still grows on the clearcut areas. The Jiuzaigou was established as a reserve in 1978 only after it had been logged, after one of China's loveliest places had been degraded. The road ends on top of a natural dam at the edge of Chang Hai, Long Sea, its blue waters extending over four miles between gray cliffs into a range of snow-streaked peaks.

Hu Jinchu and I climb up a ridge through a somber half-darkness of cedar, spruce, and fir. The descent on the other side is precipitous, the slope broken by outcrops and almost vertical pitches where the nearest handhold always seems to be a rosebush. Bamboo is scarce, and pandas are only rare visitors. Not wanting to climb back up, we make our way to the edge of Chang Hai and then follow its narrow shoreline. Cliffs drop away directly into the lake in places, but we surmount these obstacles. All except one. Only by clambering high up an avalanche gulley can we bypass the cliff without backtracking. Near the top a ten-foot gap, a slippery, almost-sheer patch of hard-packed gravel, halts our advance. With a laugh, Hu Jinchu leaps and gains the other side. Far below, I see Li Yuan, an English student from Nanchong College who has accompanied us on this trip, on a raft of three logs with a local forester. They seem to be not so much floating as suspended between the blue of lake and sky, hovering over the rippling reflections of the peaks. Rather than follow Hu Jinchu's valiant lead, I hail Li Yuan and am transported around the cliff, all of us paddling laboriously with wooden slats.

6 May. We trail up a well-worn livestock path to the west of the Rizegou until forest gives way to alpine grassland. Five blue eared pheasants, plump blue-gray birds, flush from a thicket with a clatter of wings and nasal cackles. Herdsmen use these pastures heavily in summer and still burn forest to increase grazing land, but this early in the season we have the uplands to ourselves. We wander on over meadows and through stands of cedar. There are old takin tracks, but no sign of deer, though earlier I found a cast antler that Hu Jinchu identified as that of a white-lipped deer, a species endemic to the eastern Tibetan highlands. At almost twelve thousand feet, near the last grove of trees, we take shelter in a herdsman's lean-to for a frugal lunch of crackers and hard-boiled egg while a snowstorm erases the spring. Then we hurriedly descend in cloud, the sun casting shafts of pale light; I leave the harsh beauty of these uplands with regret. In the forest we come upon another lean-to, this one a poacher's camp. In it are several armloads of coiled wire snares. These we bury beneath rock and moss.

Bi Fengzhou of the Chengdu Forest Bureau has arrived, and with him are Ken Searle, director of Hong Kong's zoo, and his wife, Sue, and Mary Ketterer from WWF–Hong Kong. We have a convivial evening crowded around the warmth of a stove. Hu Jinchu, as is his wont on such occasions, tells historical anecdotes about pandas, his face glowing with animation and good cheer.

During the Tang Dynasty early in the seventh century, he relates, Emperor Tang Taicong once held a grand banquet for those of his subjects who merited awards. The reward was a panda skin. Fourteen were given. There were two brothers, Xue Wan Jun, the older, who had died, and Xue Wan Ze, who was at the banquet. The emperor first called upon the older brother. Everyone was startled because he called upon the dead. But the emperor said, "Although he died, we must remember his meritorious service. It is the custom of China to burn our offering so his soul will receive it." And a panda skin was burned.

8 May. Hu Jinchu and I walk up a dry creek bed just east of headquarters. Ahead limestone peaks tower awash in light, fossils from an ancient sea frozen in rock. Higher up, the creek holds water, and as the tree canopy closes above us, the resonance of gurgling and splashing deepens. Then we hear a panda call, hoots and squeals and barks not far ahead. We decide to wait, hoping to find out what the panda will do. I recline in a shaft of sun, a log pillowing my head, and watch the flow of light and dark on the foliage. The panda calls several times more, then becomes silent. I doze. After lingering for more than three hours, we decide to investigate the site from which the panda had called. There is a fir, its bark heavily clawed. Droppings litter the area, and well-beaten paths lead to the creek. The panda has been around for several days, probably trying to entice a female to the spot.

We retrace our route. Something tugs at my foot and holds me; I am caught in a poacher's snare.

12 May. I am glad that Qin Zisheng could join Hu Jinchu, Xiao Qiu, and myself on our trip north to the Tangjiahe and Wanglang reserves in the Min Mountains. Qin is the botanist with our panda project; a compact person in her fifties, round-faced, her hair in a severe bob, she is a diligent worker and most knowledgeable about the rich flora of the region. Although the foreign botanists with the project have found her crusty and quite secretive about her work, as well as unwilling to share, she is always most helpful when I need her assistance. And on trips she is like a cheerful chipmunk, singing and chatting and making the long hours on the road and in camp pass pleasantly. Tangjiahe, unlike Jiuzhaigou, is said to have many pandas. Because it is a prime site being considered for another research base, we have budgeted one week this May for our survey of it, and it is with more than casual interest that I look at the forested ridges on entering the reserve.

From Maoxiangba, the reserve headquarters, we explore several valleys on successive days, all logged to varying extents and now overgrown with saplings, brambles, and vines almost tropical in their luxuriance. If there is a trail, walking is pleasant in spite of the heat and humidity, but otherwise it is the kind of dense and thorny vegetation only a takin could love.

Tucked into a bend of the road is a ramshackle hut, once occupied by a timber crew. With a bit of renovation, such as a roof, it could make a good research base. Pandas roam the surrounding hills, we have seen golden monkeys, and takin are fairly common. Perhaps we will return.

Southwest of Tangjiahe lies the county town of Pingwu. An earthquake destroyed part of the town in 1974, but a small temple survived unharmed. It is one of the loveliest temples I have seen, with red columns and green-and-blue brackets supporting the tiled roof; inside is a thousand-armed Buddha. The Temple of Royal Benefaction was built in the fifteenth century by provincial governor Wang Shi as a replica of a temple in Beijing. But this act worried the emperor, who thought Wang Shi had aspirations to the throne. A representative of the emperor therefore visited Pingwu to gauge the governor's intentions. Although the governor showed his subservience by erecting a tablet proclaiming "Long, long live the emperor," the emperor also erected a tablet, borne by a tortoise, bearing an edict forbidding the construction of any more temples in the town.

From Pingwu, a road penetrates north to the Wanglang Reserve, following the Fajiang River on whose mud-gray waters men pole rafts of logs toward the lowlands, guiding them through rapids and around turbulent bends. Where the valley widens, we enter a region occupied by the Di, a Tibetan people who wear a distinctive flat, white felt cap with one to three long rooster feathers bobbing jauntily. The fields of the Di give way to scrub and forest, and at an elevation of about eight thousand feet, we enter Wanglang. This reserve, established in 1965, consists of some hundred square miles of mountains contiguous with the Jiuzaigou reserve to the west. Many pandas were lost in the mid-1970s, the result of a massive bamboo die-off, and only an estimated ten to twenty now survive in Wanglang. Some animals starved and some may have emigrated, perhaps spurred to leave by an earthquake in 1976 when the mountains bounded like frightened musk deer, causing forests to topple and cliffs to crumble, and avalanches spewed earth, rock, and trees into the valleys.

Parts of the reserve have been logged, but forests in two major valleys toward the west are virtually intact, and we want to check these for wildlife. At the mouth of the furthest valley, wild peonies bloom and a spangled drongo, its plumage velvet black, passes by. The forest is dark and clouds dim the peaks. I walk ahead of the others, the murmur of a stream on one side, my tread so silent on the carpet of conifer needles that when a twig snaps underfoot the piercing interruption startles white-tailed leaf-warblers from nearby willows. Most bamboo stands are dead, the tall stems like upended spear shafts. Continuing up-valley under gray cliffs, I come upon a glade covered with tiny pink-flowered rhododendron shrubs. Takin have rested here and so has a leopard, judging by spoor, and I do too, waiting for the others to catch up. I am at ten thousand feet, and snow-capped peaks above have a halo of cloud.

For the next two days it rains, and fog swirls around us as we stumble and slide along mud trails. Once we push through the tangled growth of a valley still verdant with bamboo until we emerge disheveled, soaked, and cold, yet happy with a find of fresh panda droppings. Rocks, streams, meadows, and forests mold a person's mind, just as they shape a panda's habits, and this reserve, more than Wolong, Tangjiahe, or others, reminds me of parts of Alaska where I lived for four years; it gives me a feeling of coming home. But too few pandas remain to justify a study here.

26 May. We return to Pingwu. It is my birthday, my fiftieth, a special occasion in China, for at that age a life's work is traditionally done, one is old, and it is acceptable to grow a beard. County Commissioner Wei Zhangjun, whose forceful way of speaking makes all his comments resemble orders, has a special banquet for me. With expansive goodwill and the Chinese ability to make an occasion memorable, he stuffs us with innumerable courses, including a cake emblazoned with the character for long life.

10 June. We are southwest of Chengdu at the town of Leshan, famous for a huge sitting Buddha two hundred and thirty feet tall that is carved into a cliff at the junction of the Min and Dadu rivers. Unlike the sparkling creek at its headwaters, the Min here is broad and silt-laden, and during floods it may inundate the countryside. To protect the inhabitants, monks in 613 A.D. began to carve the Buddha statue, a task that required ninety years for completion. There may be beauty in mere size, but this statue has a lamentable lack of aesthetic value, the face without serenity, the body without grace.

We cross the swollen Min River on a ferry, and then our road wanders for several hours among cultivated hills until we reach the town of Mabian. As I settle into my room at the hostel, about a hundred students from a nearby school, smiling and giggling, mill around the door and peer in the window for a view of the foreigner; I feel like the inmate of a zoo. After dinner, Hu Jinchu, Qin Zisheng, and driver Li, and I go for a stroll, as we do on most evenings during our travels, down a narrow street of wooden buildings packed together. Clusters of red peppers hang under the eaves, old people sit on doorsteps chatting, and children skip and race around. We continue past garden plots of green peppers, eggplant, and beans. Smoke filters through the tiled roofs of the peasants' huts, and just beyond, dark ridges are piled on top of each other. Beneath a large fig tree we stop to look over the murky Mabian River swinging swiftly past town. Boys amble home carrying bamboo poles and strings of the small, pink fish they have caught. Hidden in the turbid waters are giant salamanders, *Megalobatrachus davidianus,* another species named for Père David. They are enormous amphibians weighing up to a hundred pounds, gray-black in color, with gargoyle mouths and minute, beady eyes. Considered a delicacy in Chinese cuisine, they are now rare.

Mabian lies at the edge of Yi territory; the Yi number about five million, and their homeland extends across southern Sichuan, western Yunnan, and western Guizhou. The cultural affinity of this tribe lies with Burma rather than with China. Although Yi and Han now live peacefully side by side, such amity is recent, as a 1936 report notes.

> In Mapien [Mabian] and Opien—Northern frontier towns—there has been frequent rebellion against Chinese rule on the borders. After the rebellion has been quelled, the heads of the various clans of Lolo [Yi] are called together on some large public space, and there they go through the ceremony of reconciliation. An ox is killed as an admission of fault and the blood is sprinkled in various places. The skin is then stretched out in full view of both parties, and vows of fealty are pledged over it.

Ancient hostility and fear still linger between the two cultures. "Don't cry or the Yi will come and get you," Chinese parents in these parts admonish their children. Today as we drive toward Mabian, headed for the Dafending Reserve, Hu Jinchu regales us with a tale of the fate of a recent Chinese expedition in Yi country during which two members mysteriously vanished. And the fate of J. W. Brooke, a British surveyor, who was killed in 1908 by the Yi southwest of Mabian, is often retold in old accounts of this region. A disagreement arose between Brooke and a local chief. Brooke put his hand in a friendly gesture on the chief's shoulder but received a fierce cut with a sword in return. Having shot and killed the chief, Brooke retreated to the roof of a hut where, besieged by the Yi, he was knocked insensible with a rock and then slain.

11 June. At the Dayuenze forest station, Jilin Zhiha, the Yi who worked with us at Wuyipeng, greets us. He is resplendent in a red turban, white blouse embroidered with a black design, and baggy green pants; an amber bead dangles from his left ear. Everything has been arranged for our arrival, and guides and porters are ready for a four-day trip on foot into the panda reserve. A Yi narrative poem begins, "The forebears of the Yi tribe, opened the high mountain wilderness." Indeed they have opened it so thoroughly that from the forest station we can see only cultivated hills except to the west and northwest, where a high wooded massif, the Dafending Reserve in the Liang Mountains, is barely visible through a veil of haze.

By noon we are on our way, a caravan of fifteen people winding single file among fields of maize and beans toward a long spur that will lead us into the forest beyond. It is humid and hot, and not a breeze stirs. When the porters stop to rest, I forage along the wayside for wild strawberries. After four hours we reach the forest's edge at an elevation of about five thousand feet. There is shade now, but the trail is awful, mud churned by hooves; livestock has been driven up through the forest to graze on alpine pastures. The forest has been damaged by intermittent logging, but we pass dogwood in bloom, and white-bellied pigeons flap noisily away. I examine several bamboos new to me, among them *Qiongzhuea,* with its greatly bulging nodes, a species used in the manufacture of walking sticks. By the time we reach an elevation of seventy-three hundred feet, we have passed through the ranges of five bamboo species, each growing in a narrow belt on the mountain. (Thirteen species are said to occur in this region.) This diversity of bamboo assures the panda a food supply even if some species become unavailable during periodic mass flowering and die-offs. It is now dusk, and for the first time on this mountainside we have found water—several stagnant pools—and hurriedly establish camp. Beneath a rhododendron laden with moss, I spread my sleeping bag on a mattress of bamboo branches and protect it from rain with a tent fly. Fortunately I have brought a mosquito net, for the tiny blackflies here are fierce. Waiting for dinner, we squat by the fire. The dry bamboo stems we use for fuel explode like gunshots, and a gray nightjar calls in the darkness.

12 June. We spend the day crisscrossing the forest in search of panda spoor. Anyo Dago, a tall, handsome Yi in his twenties, leads us with quiet efficiency, and his friend Jilin Zhiha acts as translator between Yi and Mandarin, though now, back among his own people, he is both more formal and subdued than he was at Wuyipeng. The two Yi have a close bond, for Jilin Zhiha's seven-year-old son is betrothed to Anyo Dago's small daughter. The weather is moody, with clouds low on the ridge in the mornings and rain and fog in the afternoons. We find evidence of pandas, but spoor is less abundant than we expect and the Yi agree that the animals, or *otche,* as they call them, are scarce. Local estimates of panda numbers in the one-hundred-and-fifteen-square-mile reserve vary from twenty to one hundred, figures that lack authority. We find no tracks or droppings of musk deer, takin, leopard, or any of the other larger mammals, except red panda, and wonder if hunting has virtually eliminated wildlife here.

At lunch time, Anyo Dago makes a fire and then gathers an armful of *Chimonobambusa* shoots, green and tall. He throws the bamboo shoots into the fire where they blacken. Several minutes later we fish the shoots out, peel off the charred sheaths, and eat the crisp cores that taste like asparagus. We agree with the Tang Dynasty poet, Bai Juyi:

> I have long lived in Chang'an and Luoyang, yet
> I never could get enough bamboo
> shoots to eat; here now there are many;
> do not wait to eat them, for soon the south wind
> will rise and they will grow into tall bamboos.

16 June. A side road takes us toward the northern end of the reserve where we meet five porters. Once more we climb through hills on trails muddy from incessant rain until cultivation gives way to shrub. On reaching the crest of a ridge, we find only forest before us; off to one side is the cone-shaped Jigung Shan, Rooster Peak. The trail plunges into a valley and follows the banks of a stream among massive trees, their trunks encrusted with moss, their leafy canopies closing over a hundred feet above our heads. There are *Lithocarpos* and *Castanopis*—both members of the beech family—and other broad-leafed evergreen trees, among them the dove tree, usually rare but here lining the stream. Bamboo grows exuberantly as always in clearings, but the deep shade of the forest imposes restraint, and stems there are widely spaced. The air is still and waterlogged; cicadas screech. It is the finest stand of trees I have seen in Sichuan, tall and stately like tropical rain forest, one of the last virgin tracts of a type that once covered the vast lowlands of the Red Basin and surrounding hills. Yet these last forest giants are being felled illegally by the Yi to make roof shingles, piles of which are drying on racks along our route.

Leeches thrive among the moldering leaves in such numbers that we dare not camp on the forest floor and retreat instead onto a gravel bar in the stream, where there is just enough room to crowd our tents.

17 June. For breakfast the Yi bake a flat buckwheat bread on the coals of one fire, while the Chinese boil rice for themselves over another fire. Taking my bowl of rice, I join the Yi and trade it for bread, crusty and hot.

Shafts of sun penetrate our camp as we leave to climb Jigung Shan. Leeches attack in impressive numbers as soon as we leave the safety of the gravel bar, clutching our pants and shoes until we flick them off, but a few hundred feet uphill, away from the humid warmth of the valley bottom, they vanish. Farther up, beyond the tall forest, we ascend a limestone ridge covered with twisted rhododendron and bamboo; there are a few old panda droppings. At an elevation of ninety-four hundred feet we stop for a lunch of peanuts and several wild onions we have picked. For once the mountains are clear of cloud. The highest part of the reserve reaches an elevation of over thirteen thousand feet, and in the distance I can see alpine grasslands above a dark belt of fir. Somewhere to the west, near the village of Yehli, the Roosevelt brothers shot their panda in 1929.

18 June. It rains much of the night and it still drizzles in the morning as we climb yet another high slope. Muddy and wet, we return to camp in midafternoon. I amble alone over a nearby hillside among the big trees to measure their size and just to be close to them, silent under the vaulted canopy. Later I hear shouts, muffled by the forest as if from a great distance, and I realize that it is time for dinner. My last day in the field on this survey is over.

Within four days of completing the survey, I was on a flight to Beijing where for several weeks I would join Pan Wenshi, Qiu Mingjiang, and others to finish preparing a report on our research to date. As the plane droned northward over the unrelieved monotony of villages and manicured fields, I thought of the magnificent forest of Dafending and the startling transition to denuded hills at the reserve's edge. It was a powerful reminder of what I had observed again and again during our survey. Sichuan's forests, the second largest storehouse of trees in the country, are rapidly disappearing.

Such a loss of forests means death to the panda. Already panda populations are small and isolated, confined to high ridges and hemmed in by cultivation. The survival of any species may ultimately depend on its genetic diversity. With time a small population will lose this diversity through inbreeding. In the short term, inbreeding may have an impact on reproduction by reducing fertility and the viability of the young; in the long term, a population needs genetic variation to be able to adapt to changing environmental conditions. To prevent an erosion of the panda's evolutionary potential, populations should be large, preferably at least several hundred animals, something not now possible in the fragmented habitat. But as I noted in a report to the Ministry of Forestry after our survey, it was still possible to expand existing reserves to protect more animals and allow populations to increase in size. For example, in the Min Mountains, the Baishuijiang, Tangjiahe, Jiuzhaigou, and Wanglang reserves could be combined into one large reserve with only a few square miles of additional terrain to connect them. This expanded protected area would prevent the pandas from becoming isolated in four separate reserves when the intervening forests are logged, as they surely will be soon. It is also possible to plant short corridors of bamboo that would enable animals to travel between nearby but isolated habitats. Without such measures, some of the small populations of pandas will inevitably become extinct unless they are saved through an intensive management program. In theory such a program could provide genetic diversity by actually shifting a panda every few years from one population to another, but in practice the techniques of successfully introducing new animals into an area remain virtually unknown.

Although I considered such matters in my concern for the panda's future, they lacked urgency; genetic deterioration is a slow process in a leisurely reproducing species like the panda. If what I had seen in Wolong and Jiuzhaigou was indicative, then poachers would eliminate the panda long before inbreeding could become a problem. And if any animals elude the poachers, they might live only long enough to see the last forests logged. Converted to farms, felled in large-scale government timber operations, cut illegally by local people, burned to increase pasture, the forests of Sichuan decreased by thirty percent between 1949 and 1980, and the rapid destruction continues. The massive afforestation campaigns that over the years have received so much publicity in China have had relatively little impact in most places. Two-thirds of all planted trees soon die of neglect and poor management. As a Chinese saying goes, "Trees everywhere in spring, just half left by summer, no care taken in autumn, all trees gone by winter."

In 1979, a national conference on forestry in China produced a rather gloomy forecast: "according to the estimate based upon the actual annual rate of reduction, by the end of this century there will be no trees to harvest." Only thirteen percent of China now has forest cover, and of this a mere four percent is mature, productive timber.

The loss of forests has caused environmental degradation on an immense scale, significant far beyond the threat to the giant panda. With tree cover gone, the land is unable to retain water, and the result is severe floods, reduction of stream flow during dry seasons, and droughts, erosion, the advance of deserts, siltation of dams and waterways, and the destruction of entire ecosystems with their many plant and animal species. In 1981, Sichuan was hit by terrible floods that affected ten million people, and the government blamed the disaster on watersheds denuded of trees. There has been a pointless waste of natural wealth that the country will need desperately in generations to come. Once China's forests provided the genetic prototypes of the domestic orange, walnut, apple, cucumber, and tea plant. And there will surely be new discoveries as there have been in the recent past. In 1941, one of this century's great botanical discoveries was made when the metasequoia, a superb tree thought to have become extinct one hundred million year ago, emerged alive on the Sichuan-Hubei border. The lovely silver fir was found in Guizhou in 1950.

China's current conservation problems are largely the result of irrational and destructive policies initiated by Mao Zedong during the 1950s and 1960s. Mao Zedong was this century's greatest revolutionary leader. After centuries of misgovernment, civil wars, and corruption, followed by a hundred years of oppression and interference by foreign governments that robbed China of sovereign rights, annexed territory, forced the country

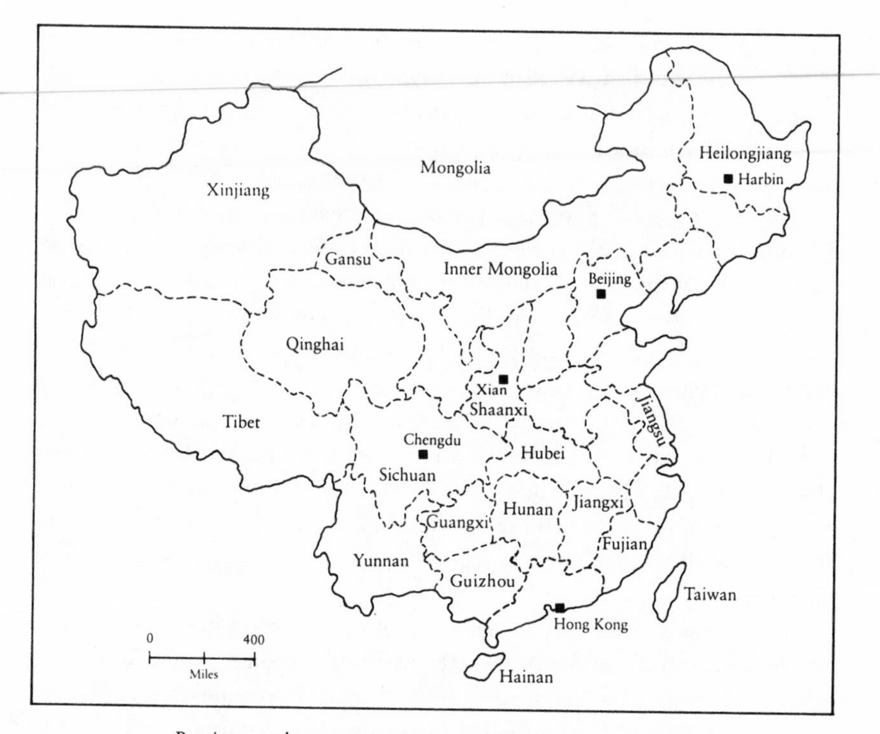

Provinces and autonomous regions of China mentioned in text

open to trade, and introduced traffic in opium, Mao Zedong unified his vast country. He gave it purpose and pride when on the first of October 1949 he spoke from atop the Gate of Heavenly Peace in Beijing: "The Chinese people have stood up . . . nobody again will insult us." Mao Zedong's vision was grand, his world impact immense. Yet his achievements are seriously flawed in part because he lacked ecological awareness; his Four Errors cloud China's potential for a prosperous future. His lack of ecological sensitivity is surprising, for he was born into a family of well-off peasants in the remote mountain village of Shaoshan in Hunan Province. As a youth he carried water from the well and tended ducks and buffalo. His major error, the one with the greatest long-term impact, was to encourage unlimited population growth: "More people mean greater ferment of ideas, more enthusiasm and more energy." During the first three decades of Communist rule, China's population increased by four hundred and fifty million people, reaching one billion in 1982. The population now stands at close to 1.2 billion, a figure that in spite of current efforts to control population growth will continue to undermine efforts to improve health, education, industry, and agriculture.

In 1958, Mao Zedong demanded a Great Leap Forward in socialist production, his second error. Peasants across the country were urged to smelt iron and steel in their backyards. Millions of trees were cut to produce charcoal for the blast furnaces; tons upon tons of iron were manufactured, then discarded as unusable because of impurities. Crops in the fields were left to rot while people smelted. The government believed its own inflated agricultural harvest statistics until famine struck. By 1961, when "economic readjustments" had been made, the famine had claimed the lives of an estimated twenty to thirty million people. Per capita grain production was at the same level as two thousand years ago. Animals were slaughtered in incredible numbers for food, depleting one of the world's great wildlife populations. Hardly were these bitter years over when in 1964 Mao Zedong made a third ecological error by urging China to "take grain as the key link." Fragile pasturelands were plowed to plant grain, and the soil was given to the wind; woodlots on steep hills were felled to sow grain, causing erosion; and even fruit trees were chopped down by people who feared being labeled "counter-revolutionary" for growing something other than grain. Fortunately this policy was abandoned within two years. But the chaos of the Cultural Revolution—the fourth error—produced further environmental neglect and destruction, as well as great human tragedy, from 1966 to 1976.

I can easily understand that people who are poor and have suffered for decades, who for the first time in their existence have the opportunity to create a comfortable life for themselves, are generally not receptive to conservation. A farmer's goal, as quoted by Vaclav Smil in his book *The Bad Earth,* is to use everything, and to use it now:

> If there's a mountain, we'll cover it with wheat.
> If there's water to be found, we'll use it all to plant rice.

Yet continued deforestation will have a grave impact on the development of China. To travel through the mountains of Sichuan creates concern for more than the future of the giant panda.

10

Zhen-Zhen Eats Bitterness

Unlike male pandas, who readily share their home ranges with others, females are more exclusive, and their ranges are well dispersed with little or no overlap. Years ago, Zhen had selected the crest of the ridge above Wuyipeng as her home. Except for an annual spring excursion downhill to feast on bamboo shoots, she had stayed there in seclusion, her peace only occasionally broken by a woodcutter, hunter, or collector of medicinal herbs. Then, in April 1978, the Wuyipeng research camp was established within her range, and trails were cut through the expanses of bamboo. Human voices became common, and strange new odors from camp drifted over the slopes, but doggedly Zhen stayed in the area she knew and liked best. As the only panda in the vicinity of camp, she naturally became a focus of attention and a source of valuable information. She adapted well to this human invasion, an innocent creature unable to foresee that, by being so malleable, a fate beyond her control would one day determine her future. Not that she often revealed herself—even though I hiked the trails daily I saw a panda on an average of only once a month—but with her nose and ears, Zhen could monitor our doings.

In May 1979, tracks of a female, presumably Zhen, with a young were seen above Erdaoping. Following her apparent maternal success that year, Zhen failed to rear an offspring in either of the following two years. But we had high hopes for her in 1982. Although we did not observe her mate, Wei and at least one other male were near her that spring. I left Wuyipeng in midyear of 1982 to write a report on our research and survey other panda reserves, and after that my news of Zhen was like contact with a distant friend: a little gossip from others, a Christmas card, an occasional get-together.

1982. The Wuyipeng staff told me about Zhen's autumn activities. She was radio-located on 28 August near a den tree about a quarter mile from where she had had her young the previous year. Two days later, she had deposited five twigs in the hollow of the tree as if to construct a nest, an indication that she might soon give birth. On both the first and second of September she was observed in the den, which she did not leave even when she was disturbed by several people, and on 7 September the squawk of an infant was heard. Ten days later her radio signal remained inactive and, thinking that something had happened to Zhen, the staff investigated. She sat motionless in the den with her back to the entrance. Shouts from a distance of fifteen feet and even a thoughtless poke with a stick elicited no response except that she glanced back over her shoulder at the intruders—behavior considerably more placid than her response to Hu Jinchu and myself the year before. She was without her collar. Her infant, big as a cat, was last observed on 15 October when it was discovered by botanist Qin Zisheng in another hollow tree where Zhen had cached it. Qin Zisheng was studying bamboo near the den tree but, on hearing Zhen approach, she quickly terminated her investigations. Zhen's baby disappeared before year's end.

On 10 November, Zhen abruptly descended from Erdaoping to Wuyipeng, where the camp staff fed her goat meat. It was an unusual trip for her to make at that season, and her visit to camp was preemptory, as if she had perhaps contemplated such a step for a long time and finally overcome any reservations. Within twenty-four hours she was back at Erdaoping.

1983. Zhen was neither recollared that winter nor during the whole of 1983. Without a radio signal to locate her, we had only glimpses of her life. A panda was observed in a tree at Erdaoping on 28 April, with four males including Pi and Wei waiting below, good evidence that the animal in the tree was a female in heat. Pandas are difficult to identify individually by their coat pattern, and without her collar the observers could only infer that the female was Zhen.

That spring she made her usual excursion into umbrella bamboo to eat shoots, coming near camp several times, always without an infant. During one night of heavy rain, after I had gone to bed, Xiao Qiu came to my tent. "Are you asleep. Zhen-Zhen is by our tent." She had snuffled through kitchen garbage, and when I arrived she was eating shoots. The weak beam of my flashlight revealed what appeared to be a gray boulder with diamond eyes. The night was black as a panda's ears, and Zhen used her sense of smell rather than sight to find shoots, walking slowly, her muzzle moving low among the stems until she discovered a shoot, which she then ate in her usual manner.

I was in Wolong for a two-week stay in mid-December 1983 when the staff informed me that they had seen a year-old young with Zhen on three occasions during September. When she was trapped for recollaring on 4 December, Zhen was lactating. (Zhen promptly succeeded in removing this collar.) Everyone thought that her 1982 infant must have survived after all. But on turning the evidence over in my mind, I reluctantly disagreed: we had not seen a baby with her that spring, there had been no small droppings in her beds, and a female usually does not suckle an offspring for well over a year. I deduced that Zhen had given birth that year in late August or early September but that the newborn must have died almost immediately. She then must have adopted an orphan at least temporarily until her strong maternal urges subsided. Where would she find such a timely stray? I fervently hoped that the young whose mother had been killed by a poacher the previous April had actually survived to find warmth and protection from Zhen.

During my return that December, I had stopped at Hetauping. Construction of the research center was almost completed, and the pandas from Yingxionggou had already been moved into their new quarters; in these more comfortable surroundings we hoped they might breed. I noted a new male. His name was Hua-Hua, I was told; he had been caught in our study area in late November. Why had he been made a captive instead of being collared and released? Who was responsible for such a deplorable decision? Once we had planned to make a long-term study of an undisturbed panda society, but the population had been disrupted by poachers. And now this. My questions were answered with the kind of straightforward ambiguities and evasions that underrated the value of plausibility. I suspected another reason. Some months before at a meeting, Hu Jinchu had indicated that eight pandas would be caught to fill the research center. I voiced strong opposition to this idea, one whose time, I hoped, would never come. Had it arrived this soon? My next step was obviously to fight for the release of Hua-Hua.

1984. In mid-1983, Hu Jinchu and I had evaluated the research done at Wuyipeng during our long absences in other panda areas. We agreed that there were various unhealthy tendencies in the work and that these must be dealt with resolutely. Several recent college graduates had been added to the project to lend it scientific credibility. With less than five percent of the population attending college, a graduate represents an elite, even though he or she is often woefully ignorant of anything except what is memorized from textbooks. As in the time of the emperors, when aspiring scholars had to pass a national examination, the *jinshi,* given once every three years, emphasis still remains on memorizing. Students are not trained to think but to learn by rote, to be passive. This passivity is often coupled with a lack of interest in both profession and work place, both generally selected not by the students themselves but by superiors. And some also feel complacent about being a graduate and show contempt for workers who are not. These traits combined to create the worst possible situation at Wuyipeng. Workers like Tian Zhixiang, Zhang Xianti, and Xiao Wang, who for years had suffered the hardships of camp to collect information, were now looked down upon by newcomers who often used insolence to conceal incompetence, inertia, and a starved imagination. While there had been periodic morale problems before, relations among all of us had been cordial. Not so now among the Chinese—to the detriment of all research.

Zhou Shoude had been transferred, and his place as camp leader was taken by a new graduate, Wang Pengyan, a slight, seemingly meek man. He was a disastrous choice, for the gap between his ability and his position was enormous. Traps went unchecked for days, endangering the lives of pandas, and radio tracking was done haphazardly. Research notes, which had always been accessible to everyone, were now locked away in a cabinet. When I wanted to see them, I was told, "The man with the key is gone." A common refrain in China—the man with the key is gone. China may have an open-door policy, but it remains a land of locks—on doors, desks, bookcases, cabinets, everything. And the person with the key—and there is usually just one key—does not delegate responsibility, not even giving the key to someone else when going away for a month. A key is a credential of power. The sentence is also used as an excuse to prevent access. And, metaphorically speaking, it permeates society. To avoid making a decision or taking responsibility, it is always someone else, preferably far away, on whom the problem devolves. Wang Pengyan was eventually replaced, but it was a case of treading the old path in new shoes. The new leader, another recent graduate, was Zhang Heming, a friend of Wang's. Under their crafty coalition of misapplied intelligence and small-minded conspiracies, the study in the Choushuigou came close to disaster. As Wang Menghu, who had an appropriate proverb for every occasion, would say, "When the tiger is away from the mountain, the monkey proclaims himself king."

Hu Jinchu had asked me to find Westerners to take my place at Wuyipeng while I was busy elsewhere on the project. I found two men, both, it turned out, excellent choices. Kenneth Johnson had worked for several years with Michael Pelton at the University of Tennessee on the biology of black bear. Highly experienced in immobilizing bears and radio tracking them, he was a great asset to the program. Also needing a botanist, I telephoned Thomas Veblen at the University of Colorado, who had studied bamboo in Chile, and he recommended one of his graduate students, Alan Taylor, who had a special interest in the dynamics of forests, ideal for the research on the interactions of bamboo and tree growth we had planned.

I went home in mid-January 1984 for one month to celebrate a belated Christmas with Kay and to deal with various conservation matters at my New York Zoological Society office before returning to China for the rest of the year. In February 1984, I met Alan Taylor and Ken Johnson in New York, and the three of us then proceeded to Wolong, where I planned to spend ten days introducing them to the pandas and staff. I would then go north to establish a second research base in the Tangjiahe Reserve.

The news after my month's absence was unremittingly bad. The baby panda who had attached itself to Zhen had been captured and taken to the research center. There were several versions of how and why this had been done, some, it seemed, not inhibited by fact. In brief, I gathered that Zhang Heming had seen the youngster in a tree, climbed up and grabbed it, getting his arms scratched in the process. The baby was then left in a log trap for a night until Party Secretary Lai Binghui gave permission to remove it to the research center. I was furious, though I did my best to suppress this response. Had their conscience been swallowed by a dog? After the debacle with Hua-Hua, I had stressed that our study animals, as well as others in the reserve, must not be taken into captivity. But none are so deaf as those who do not want to hear. Hua had not yet been released. Would that panda palace, as the research center was called, become nothing but a panda prison for Wolong's animals? A home for the living dead? The newly captured infant soon became ill at the research center, and the two veterinarians there, who knew little about treating animals, sent it to Chengdu, where it died. I felt a growing weight of responsibility, helpless to overcome obstacles. In spite of three years with the project, I had been unable to develop among my Chinese colleagues a feeling for the major goal of our joint venture: the panda must be assured its freedom and survival in the wild.

A few days before in Beijing, Wang Menghu had told me of a female that had been collared by the Wuyipeng staff on 4 January. Bei-Bei, as she was called, had all but moved into camp and was now being fed there. He wondered if perhaps she should be made a tourist attraction. Upon my arrival in Wolong, I was warned not to sleep in my usual tent because Bei-Bei used it. Who was this Bei-Bei? Surely it must be Zhen, who was already familiar with camp and its delicacies.

As soon as Alan, Ken, and I arrived at Wuyipeng we were regaled with tales of Bei and given a tour. In Kay's and my tent there were about 15 droppings on our bed and a big pile of droppings in one corner where Bei had whiled away the hours; she had also chewed a corner off our desk. In the camp kitchen, I was shown how she had climbed up on a cupboard to reach a twenty-five-pound slab of pork hanging from the rafters; in the research tent, she often rested on a soft pile of bamboo samples that no one had bothered to rescue. Bei had smashed the windows of a new bunkhouse built to accommodate the staff, so now the windows were boarded up to keep her out, and the interior was like a cave, with lanterns burning even in daytime.

Zhen had several teeth missing in her lower jaw but no one had checked the jaw of the so-called Bei when she was immobilized for radio collaring. Later Ken confirmed that indeed Bei was Zhen. But the name Bei continued to be used, a bitter end for Zhen, who had achieved world renown only to end her days under the anonymity of an alias.

While the panda has been made a symbol of the fraternity of living things and there may be a certain cachet in sharing a bed with such an animal, I elected against cohabitation with Zhen. She had, admittedly, been placid on most occasions when we met, but I had little idea how to defuse her more assertive moments. I moved into the bunkhouse.

As I was unpacking my duffel, Zhen arrived in camp. Zhang Heming handed her a piece of sugarcane that she deftly peeled with her teeth, tearing off the tough bark, and then ate, juice dribbling from her lips. Wanting more, she lumbered toward us emitting snorts, growls, and rasps, surprisingly aggressive. As we scrambled away, someone threw her another piece of cane. Wang Pengyan now stepped close to her and, like a picador goading a bull, he poked her with a stick and hit her lightly until she became infuriated and charged him, cutting his retreating pants with a swipe of her razor claws. He scampered off, closely pursued, but suddenly he turned and handed her more cane. Braking to a stop, she grabbed and ate it. Now I knew why Zhen was so short-tempered: she had been trained to attack. It did not take an extravagant intelligence for Zhen to realize that her aggression produced gifts of food. Wang Pengyan had finally discovered an interest at Wuyipeng: he could gain attention by his antics with Zhen. But his devious opportunism had also turned Zhen from a creative nuisance into a dangerous animal. Self-assured, Zhen now ambled into the research tent as if she had always lived there. When I peered in through the window, she came and looked out, our faces only two feet apart. It was surreal, the study subject in the research tent looking out in a proprietary manner, the scientist roaming outside.

After a nap, Zhen came in one door of the communal hut while everyone else skittered out the other. She went to the fire, standing not by it but on it, her forepaws on a burning log, hot coals beneath her belly. The smoke caused her to sneeze, but the heat she ignored, her fur insulating her so well that she even failed to respond when fire singed her, turning white hair yellow. I tried to shoo her away, but it took Wang Pengyan with a stick to make her retreat. She left but was soon back. In a corner of the dirt floor was a small root cellar, a hole covered with boards, and for the next hour Zhen leisurely ate the sugarcane stored there. She stopped her meal only long enough to duck far into the hole, nothing but her rump visible, as she fished out another stalk. We occupied half of the room, she the other, a companionable sharing by the light of the fire.

Zhen's schedule was erratic, but she came to camp at least once a day to slurp down a handbasin full of rice porridge and to scavenge whatever she could. Our life revolved around her. The cook was afraid to begin breakfast until daylight for fear of encountering her in the kitchen raiding *mantou* or meat. With errant enthusiasm, Wang Pengyan and some others did nothing but monitor her signal all day, waiting to feed her. Our travels became furtive as, ever alert, we expected Zhen to appear and demand food; we even carried pieces of sugarcane to buy her favor in the event of attack. Hardy souls who stepped from the bunkhouse after dark as often as not dove hurriedly back through the door. Alan and Ken had decided to brave the nights in tents. After all, Tian Zhixiang still lived in one behind a barricaded door. One night, not long before dawn, I heard a Chinese say, "*Weiguoren,* foreigner." Waking fully, I also heard Ken yell. As I hurriedly put on my clothes, Ken burst into the bunkhouse wearing only white longjohns, unlaced boots, and a Chinese fur hat askew. Not that he owned much more—all his belongings had been stolen on the way to China when they were with a company that transfers baggage between New York's La Guardia and Kennedy airports. While Zhen tried to claw through the back of his tent, Ken shot out the door; she then entered and investigated, as was her right, we being just squatters in her home range. Everyone dozed off again, but soon Alan called for help. "Lao Tian!" Tian Zhixiang lived in a tent next door to his. Zhen was obviously determined to visit Alan. "*Meiguanxi,* never mind," Lao Tian replied helpfully. Wang Pengyan and I ran out to lure Zhen away with sugarcane while Alan made a dash for safety. Both Ken and Alan now moved into the bunkhouse, joining the ten others, giving up privacy in favor of other advantages.

Zhen, it was clear, would have to leave camp as soon as possible. I outlined the problem in a letter to Lai Binghui, using her alias Bei-Bei:

1. She has been trained by some Wuyipeng staff to demand food aggressively. She is now dangerous, and may seriously injure someone.
2. She disrupts the Wuyipeng research effort. Workers spend more time feeding Bei-Bei than doing their tasks. Tents cannot be used; people are afraid to go into the forest.
3. It is wholly wrong to transform a wild panda into a porridge-eating camp panda for no reason other than to amuse people. A natural reserve is for wild animals; if people want to see tame ones they can go to a zoo.
4. Because of Bei-Bei, Wuyipeng is turning from a research camp into a tourist camp with too many visitors, news people, and others all wanting to see her.

In principle, it would be easy to stop feeding her and chase her off. But Wang Pengyan and several others wanted her in camp, and Lai Binghui, whenever he was confronted with the dilemma of having to make a decision, sided with the majority. Another alternative would be to move Zhen to another part of the reserve, even though it would mean losing yet another study animal. In weak moments, I was admittedly ambivalent about having Zhen around camp. It was, indeed, a way to observe the resources of her intelligence as she adapted to new situations. And she certainly added spice to our daily routine. Yet her altered presence was also infinitely sad. When a creature is larger than life, you expect tragedy or at least poignancy when its fate has been changed. But Zhen had been transformed from an imposing and mysterious creature to a mere panhandler. Again and again I had emphasized that there should be one, and only one, basis for evaluating any action: what is best for the panda.

Matters had not been resolved in late February 1984 when I left for Tangjiahe. The collective irresponsibility of bringing a panda into captivity is quick and casual, the decision made at the local level, I noted sourly to myself, whereas to return an animal to the wild seems to take months of discussion among officials at Wolong and in Chengdu and Beijing. Ken and Alan now kept me informed of events at Wuyipeng, and I was sent newspaper and magazine clippings as well. Zhen (as Bei-Bei) became a media star of the isn't-she-cute variety, but not one of the many articles questioned the ethics or scientific validity of fettering her to camp with food.

The situation at Wuyipeng deteriorated further. As Ken wrote me,

"Wolong is more out of control than leaders in Chengdu and Beijing realize." As if possessed by a fox demon, Wang Pengyan, Zhang Heming, and another graduate at headquarters, Shi Junyi, became a Gang of Three whose aim, it seemed, was to destroy the research program. Hu Jinchu seldom visited Wuyipeng, and Lai Binghui remained ever the chameleon with great versatility in his convictions. The staff trapped two pandas and without informing Ken released them uncollared: more radioed pandas mean more work. In fact, the staff had held a meeting and decided research was too arduous—even though each person had to be afield at most one morning every other day. Ken then often had to check all traps on his own.

Zhen, alias Bei, became more unruly, according to letters from Alan and Ken.

> Bei-Bei got inside the big hut, apparently let in by Wang Pengyan. This almost happened again last night; sugarcane had been placed at the entrance to the doorway. The cook and I were stranded outside. . . . The week before she tore out the stove in the kitchen; another has been built. She has become more aggressive and on the night of March 13 was chasing people with gusto. . . . The workers here would love to sit around the fire and feed fat pandas; everyone believes this is the camp mission. . . . If Bei-Bei does not get removed some kind of arrangement for safe housing will have to be worked out. There are now 12 people in 10 beds in the hut. . . . I drift from place to place looking for a place to sleep.

Wang Menghu came from Beijing to settle Zhen's fate, but the Wuyipeng staff resisted outside interference. Ken sent me a letter with his impressions.

> Wang Menghu produced the feeling that he wanted us in Tangjiahe out of the way. We also got the same feeling from people here. . . . Enjoy the peace and quiet at Tangjiahe. That is a commodity in short supply here.

I did enjoy some peace, for letters from Wuyipeng took weeks to reach me, apparently held up somehow en route.

One of Zhen's "little pranks," as one newspaper termed her behavior, was to slash Wang Pengyan in the leg. After this Wang Menghu achieved a token victory: Zhen was banished but only across the ridge into the Yingxionggou, barely outside her usual range. Released there on 23 March, she strolled back home to the top of the ridge and stayed there, perhaps not keen to relive her media triumphs at Wuyipeng. But Wang Pengyan, faced with public oblivion, laid a trail of sugarcane and meat to lure Zhen back to camp. When Ken objected, he was told: "I feed Bei-Bei. Leaders told me to."

Ken's main effort at that time was devoted to recapturing Long, whose radio collar would soon cease to transmit. Long, however, was exceptionally astute about avoiding traps. The previous year, when Hu Jinchu and I were busy elsewhere, camp routine had become so lax that traps went unchecked for days. Long had been caught twice, I was told confidentially, but no one had released him. Frantic with hunger, he had torn at the trap walls until somehow he had found a way to lift the door and escape. Long remained uncooperative; he was never again recaptured. But a new female, Li-Li, was caught, and Ken had to radio collar her without any assistance from the staff. Then Zhen was caught in one of our foot-snares. Ken released her, helped by Seidoh Hino from Nippon Television and his Chinese coworker. The snare had slightly abraded Zhen's forepaw, making a half-inch skin-deep cut, a trivial injury.

Having first neglected to assist Ken, the staff now unjustly claimed that he showed incompetence in drugging and handling pandas. Zhen was said to have been immobilized for three hours, when actually she was asleep for only half an hour; Zhen was said to have a limp as a result of the cut, something no foreigner in camp could detect. Two Chinese veterinarians were brought in to treat the cut unnecessarily, and when Zhen was tranquilized for this treatment, it quite possibly interfered with her estrous cycle. There was no fire, so the staff invented smoke. A foreigner is always vulnerable to rumor, however unjust. Alan was spared much of the acrimony because bamboo arouses less passion than pandas.

Confucius noted that "a gentleman can see a question from all sides without bias," in actuality a most difficult task when cultures clash. Westerners are generally concerned with the facts of a matter; truth in their view is achieved by examining each point of disagreement—in this case drug dose or seriousness of injury—on its merits. In principle, conclusions tend to be reached by examining individual aspects of a situation rather than the whole. Chinese, by contrast, often view facts from the standpoint of a value judgment, asking themselves whether something is inherently good or bad. Facts, no matter how irrelevant or misinterpreted, may be used to buttress a perception. The pragmatic slogan "Seek truth through facts," so often repeated in China, tends to remain just a slogan. The unfortunate consequence is that neither culture understands the other. The Westerner may think the Chinese deceitful and the Chinese may accuse the Westerner of being bad. It is somewhat like the Beijing opera *Three-forked Road,* in which two brothers meet on a black night and fight, thinking they are enemies, until the light from an innkeeper's lantern makes them realize they are actually friends.

But mere misunderstandings did not account for one incident. Ken and Alan had gone to headquarters to discuss again the problem of Zhen. Lai Binghui was absent, but a meeting was held with Shi Junyi, Zhang Heming, Qin Zisheng, and several Wolong leaders. Wang Menghu had returned to Beijing. With the unanimity of a religious get-together, the Chinese faced the two foreigners in an ugly mood. Shi Junyi, ever the ankle-biter, loudly and angrily harangued Ken, accusing him of incompetence, of harming the pandas and showing an "incorrect attitude." His command of misinformation was impressive. Ken's explanations were curtly disregarded with "you do not know much about pandas," and Alan's complaints were dismissed as "imagination." The painted slogan on a cliff face by the road on the way to Wolong—Down With American Imperialism—has faded. The political exhortations and martial music that still blared at dawn from loudspeakers at headquarters in 1980 had by 1984 been replaced by *Jingle Bells* and *Red River Valley*. But the righteous irrationality and malice of Cultural Revolution tactics still lingered at that meeting; one had to fear the accusers' imaginations, not their knowledge.

I was indignant when word of this meeting finally reached me in Tangjiahe. Ken had handled the situation admirably. He wrote me, "I surprisingly maintained a calm composure throughout, felt a high level of humor underneath the whole scene, felt no need to defend myself or get into their emotional frenzy. . . ." By letter to Lai Binghui, I objected to the mistreatment of pandas and WWF personnel and stated that "it will severely tarnish the image of China if the world learns what has been happening in Wuyipeng. . . . It may be necessary to call a joint meeting with representatives from WWF, Beijing, Chengdu, and Wolong to resolve matters."

The Chinese did not want a formal confrontation with WWF. Once more Wang Menghu came to Wolong and this time produced action. In a tardy purge, Wang Pengyan was removed from Wuyipeng. But instead of a prohibition against feeding Zhen, her removal to the research center was ordered. On 4 June, through no fault of her own, Zhen entered a nightmare. She had to eat bitterness, an apt expression for those who suffered adversity in the Cultural Revolution. At least Wuyipeng settled back into a research routine once more.

Three months later, in September, I saw Zhen at the research center. She was sluggish and fat-bellied—still fed porridge with barren zeal by Wang Pengyan, whose new job this was.

At this season Zhen should have been free, caring for a newborn as she had done in previous years. Zhen's neighbor was Hua-Hua, now a captive for over nine months. Their dignity had been taken away. Beneath the surface of their present lives was an inarticulate agony that filled me less with pity than with indignation. Was Hua's future to be electroejaculated once a year instead of struggling with others of his kind for the right to mate? Was Zhen's future to be drugged again and again and artificially inseminated? Removed from their culture—their society and pattern of life—their lives would be tranquil but empty, a tragedy. I was determined to fight on their behalf, to rage and scream if necessary, and in this I actually had a quiet ally in Hu Jinchu. One must, however, be idealistic in a practical way, one must define the limits of fruitfulness. We had discussions and more discussions at several meetings—and won their release.

September 23 was an important day for Zhen, one in which I unfortunately could not participate. Bearers lifted her barred cage, suspended between two poles, to their shoulders, as though they were transporting an empress in a sedan chair. They carried her across the Pitiao River to the base of a bamboo-covered slope. She barked with annoyance and tried to claw anyone within reach, once drawing blood from a careless porter. When the door to her cage was opened, Ken told me, she sat with her back to it, but finally she walked casually out and away. Released ten miles downstream from Wuyipeng, she could reach her home by crossing several small drainages. Having retained a sense of place while in captivity, she now began a leisurely journey back.

Five weeks later, on the first of November, I received a telegram from Wolong's leader:

> Zhen-Zhen arrived at Choushuigou October 30 and arrived near Wuyipeng October 31. Perhaps she will come back to Wuyipeng again. What shall we do about her? Please telegraph your suggestions.

I replied as follows:

> I have three suggestions about Zhen. 1. She must never be given food. 2. Food in the kitchen should be stored so she cannot take it. 3. Staff should chase her away if she comes to Wuyipeng. Perhaps she will then become a wild panda in her old home again. Good luck.

During her travels, Zhen indeed reverted to a bamboo diet, and Wuyipeng ceased to attract her. If she remembered the good food there, perhaps she also remembered the harassment. Once she did come to camp, but this time no one fed her, and she did not repeat the visit. She was truly free again.

In April, there had been discussions about releasing Hua. However, John Knight, a London Zoo veterinarian who had come to assist with artificial insemination, noted that Hua was not walking normally. An X ray showed that one thigh bone had been broken in three places, as if smashed by a heavy weight, perhaps by the iron door of a carrying cage. The bones were only partially mended. Hua was well again by September; one hindleg was a little shorter than the other, but this seemed not to inconvenience him. Six days after Zhen's release, Hua was taken back up the mountain where he could become part of his old community again. That he so readily settled back into the wild after nearly a year of captivity made him a pioneer whose success, I hoped, would soon be emulated by other captives. Of his homecoming we know nothing. As he visited scent posts to announce his presence, he no doubt found out that Long was still there. Zhen had not yet returned, and in her range, vacant since June, only traces of scent lingered. Pi, the dominant male whom Hua had once carefully avoided, had failed to mark his routes for nearly two months, a fact that Hua probably noted. But he could not know that Pi was dead.

I was in Chengdu's Jin Jiang Hotel on 1 September 1984, ready to return to Tangjiahe. Ken and I had met here three days earlier; he was passing through town on his way to Wolong after spending the summer in Tangjiahe, and I was returning after a summer on the Tibetan Plateau. We had shared experiences and planned next season's work before parting early that morning.

Late in the afternoon Ken burst into my hotel room. He was excited, his Tennessee twang almost staccato. His blond hair was stiff with dried sweat, his clothes were disheveled and muddy, and he smelled like a cesspool.

Pi-Pi is dead!" he exclaimed. "I dug Pi-Pi up. I saw his body with my own two eyes."

Ken retrieved from his backpack grisly proof of his story—a thigh bone with tendrils of decayed meat, three ribs, a section of rotten intestine.

When Ken arrived at Wuyipeng, Alan had told him that when the staff had radio monitored pandas continuously from 17 to 19 August, Pi's signal had remained inactive. Zhang Heming and Shi Junyi organized a search party on 20 August and found Pi's body on the following day. The two veterinarians from the research center came to autopsy Pi, not in the laboratory but quickly in the forest. A few days later it was announced that Pi had died of an intestinal obstruction. Every effort was made to discourage WWF personnel from discovering the details concerning Pi's death. John Knight, the British veterinarian at Wolong, was not permitted to observe the autopsy, and I was not told of Pi's death by Chengdu officials. The body was buried immediately. It was a cover-up, both literally and metaphorically. How had Pi died? Alan was told that Pi had been snared.

Several days later, Ken, a number of Chinese, and I went to examine Pi's remains. The forest was gloomy, and a soft rain fell as we walked the trail past Bai Ai and descended into the valley. Pushing upstream along the bank through a tangle of undergrowth, we soon came to a small gravel bar. A paw with bones visible between strips of black hide lay on the ground and fragments of skeleton protruded from a hastily covered grave; Pi's spinal column balanced on a nearby boulder.

I gazed at the remains. All too often on this project joy had passed while sorrow and failed hopes remained. But I had little pity for the animal; one abandons pity when it is useless. Bones are not a panda, and Pi would remain in my mind as an indomitable spirit battling for supremacy against other males.

No poacher would have set snares in such a dark gully; I searched the area for drag marks or other signs that would indicate Pi had been hauled to the site. I looked for saplings recently cut, animal trails, any evidence, but found nothing to indicate that Pi had been snared. Why did several Chinese claim otherwise? Was it an attempt to create trouble for Zhang Heming and Shi Junyi in retaliation for their overconfidence and arrogant smugness? What one is told is usually just a shadow of reality. Why the secrecy surrounding Pi's death?

I left the others at the site and returned alone to Wuyipeng. In past months the tent camp had evolved into a village with four large wooden huts. Soon there would be electricity. The rugged intimacy of the old days was gone. Tian Zhixiang was there, and I asked him to show me the August radio location data and activity readings for Pi, hoping to pinpoint the day he died. "The man with the key is gone," said Tian Zhixiang. The data were unavailable. I would wait, I told him, days if necessary. In a few minutes the information was reluctantly handed over. I flipped through the daily card record. Days after Pi had definitely died, he was still recorded on occasion as being active. Careless work.

After Ken had told me about Pi's death and the uncertainty surrounding its cause, I debated how to respond. I was weary with a psychic numbness; there is a limit to the art of endurance. I did not want to become involved in wrangles with indifferent Wolong officials who lacked the knowledge and ability to fill their positions, yet I was determined not to let another incident be quietly defused. Usually when a problem cannot be ignored anymore, someone low in the hierarchy admits to error, achieving salvation by being willing "to improve his way of thinking" and "realizing the seriousness of his mistakes." Self-criticism may be a way of redemption, but its deterrent effect is dubious and any reforming effect questionable. So many mistakes had been made in 1984 that a mere repudiation of actions was not enough.

As a guest in China I could not impose changes, only apply pressure. Where to draw the line between persistence and presumption? It is a difficult mountain to cross, as the Chinese would say. I decided to ask for a joint meeting of Sichuan and Wolong forest officials.

The meeting room was crowded, the Chinese tense, uncertain what flies their open-door policy had permitted to enter. Lai Binghui began the proceedings with a litany of blame for others. The Wuyipeng staff "has incorrect ideas about collaborative research. . . . Young people should be severely criticized and given more education. . . . Some have been impolite to foreign staff." Zhen had been taken to the research center because "she was physically weak," astounding news to all who had been chased by her. Regarding Pi, "I have no intention to hide anything from Chinese and foreign experts." He then welcomed our criticisms and suggestions for improving his work. On and on. Ritual phrases.

It was my turn. I apologized that I as a foreigner offered criticism. However, I had a responsibility to unburden myself in order to help the panda. I was sick at heart and angry at the many disastrous decisions by the Wolong leadership and the constant interference in the research and in the pandas' lives. Hua, Zhen, and an infant had for no good reason been removed from our study area. At least three of our pandas had died in poacher's snares. "You say we have an emergency because the bamboo has flowered," I accused them, referring to a bamboo die-off that began the previous year and now threatened pandas with starvation in some areas outside Wolong. "I say we have an emergency because of poaching. Why is the panda so scarce in the forests? Because of snaring. And there are still no patrols or other antipoaching measures."

The expensive research center, I further noted, was filled with delicate equipment donated by WWF. Yet a year after completion there is no research, no senior staff has been brought in, workers play checkers and drink tea instead of caring for the pandas, equipment is being broken and library books are being stolen. Foreign newsmen visiting Wolong to interview WWF researchers are told that this is not possible and instead are briefed, often with erroneous information, at headquarters. Why?

For what purpose are foreigners here? Certainly not to do most of the work. We must start anew, clean house; we must have monthly meetings and each time sign an agreement of the work that is to be done; we must release Hua and Zhen. For hours I droned on, giving details to illustrate each problem in Wolong and elsewhere. Alan and Ken added their comments, as did John Knight, who noted that among other problems five of the pandas at the research center had been ill when he arrived.

No one rebutted us. All Chinese nodded in agreement when Gung Tongyang, head of conservation in Sichuan said, "We must smash the iron rice bowl" that guarantees a job and a wage no matter how poor the work. "We must educate local people and make them guards of giant panda," he continued. After he had finished his discourse (courtesy demands no interruptions), I noted that as long as the government buys musk from villagers at a high price and overseas demand for musk remains great, local people would continue snaring musk deer and inadvertently snaring giant pandas, no matter how much education they have. There must be antipoaching measures.

Lai Binghui noted that there should be a "reform meeting" for the staff to "give re-education and raise social awareness." Hu Shixiu, a forceful woman known affectionately as Mrs. Panda, added that "we should be decisive in solving problems." It was all upbeat.

When the meeting was over, I felt guilty about having been so critical. I liked many of the individuals; I did not want to cause them problems just because a few "rotten apples," as the Chinese called them, were in Wolong. Hu Jinchu tried privately to reassure me. "I'm grateful you spoke out. . . . Work was influenced by people who do not know science." When I met Wang Menghu a few days later, he said: "There is a proverb in China 'To conserve yourself is to protect the enemy'." He was discouraged that day, his ebullient manner gone. "I'm losing heart in Wolong. . . . The effort is too much for what we have achieved."

Changes in Wolong might have been merely cosmetic if a reporter from *Xinhua,* China's news agency, had not been present at the meeting. An elegant woman, she quietly took notes, and afterward circulated an internal report among the top Party leadership in Beijing.

A team from Beijing, including vice minister of forestry Dong Zhiyong, came to Wolong. The vice governor of Sichuan became involved in supervising the program. Meetings of self-criticism were held. The three main troublemakers were lightly reprimanded and given different jobs. Zhang Heming soon went overseas for study, and he became acting director of the research center after his return. Some punishment! It would have been better to heed the old Chinese saying that "When you cut grass you must pull out the roots." However, the system prescribes leniency for those who admit their mistakes. Only an interpreter was punished—for revealing too much to foreigners. Lai Binghui of course retained his position. *Xin bu shang da fu*—punishment does not reach high officials. The iron rice bowl remained unbroken. However, some new and more competent researchers were sent to Wolong, there was better cooperation for a while, and interference from headquarters diminished. And, most important, Hua and Zhen were soon released back into the wild.

1985. Zhen died on 18 April not far from Wuyipeng. She was at least thirteen years old. After being taken to the research center, her body was autopsied by John Knight and Qui Xianmeng:

> Summary of cause of death: An adult, possibly aging animal died of natural causes. The congestion of the meninges may indicate a meningitis, though no culture for a causative organism could be made. Large quantities of ascitic fluid and a lack of fat reserves indicate a long-term weight loss, the emaciation being severe by the time of death. No other significant cause of death could be ascertained.

Why had Zhen lost so much weight? If it was meningitis, had she carried seeds of death from the research center? Hua had readapted well to a life of freedom.

Han, Ning, Wei, and Pi had all died, and now Zhen; only the memory of their footsteps rustled mournfully on the slopes. Of those pandas whose lives we had traced, only Long remained. Zhen had provided the Choushuigou with a special enchantment; a lovely creature in her own right, she had also become the symbol of our project. In spite of seeing her world disappear, Zhen had clung to her ridge, stoically watching the forest shrink tree by tree, observing other pandas killed by snares. Finally she herself had vanished into captivity only to reappear, her moment in time not quite over. Now she, too, was gone, but she remains with me as an insistent and heroic presence whose troubled life contains the message that the panda can indeed prevail.

11

Tangjiahe

February 1984 to January 1985

Heading straight north across the Chengdu basin, the road was narrow and swarming with traffic, crowded buses, senescent trucks belching greasy smoke, and small, noisy tractors pulling wagons. Bicycles in a steady stream veered among the vehicles and around carts pulled by listless donkeys. Almost every bicycle was burdened with sacks of grain, poles, baskets filled with produce, chickens suspended by their feet from the handlebars, or pigs trussed upside down behind the unsteady rider. Beyond double rows of poplars that edged the road, scattered farmyards were surrounded by stands of tall bamboo and fields of sprouting wheat and rape seed, a few stems in bright yellow bloom. I was tight with tension as our driver, Lu Daiming, hurtled along with a series of accelerations and sudden stops, more confident than I that nothing would stray into our path. Our new car—a Toyota Land Cruiser donated by World Wildlife Fund-Japan—was crammed with equipment and belongings, everything I would need for the new project in Tangjiahe.

It was 29 February 1984, over three years since the panda project began in Wolong. Now we would establish a second program in an area with less rugged terrain than at Wolong, and one with different bamboo species, some of which had flowered in the mid-1970s, to give us an idea of how pandas had adjusted to these conditions. I also sought an area in which pandas and Asiatic black bear inhabit the same forests. These two species resemble each other in size and build, and I hoped that an understanding of the bears would place some of the panda's peculiarities in clearer perspective.

The previous year we had surveyed several panda reserves to look for a new study site, and Tangjiahe in the Min Mountains of northern Sichuan bordering Gansu Province had attracted me immediately. Compared with the confining valleys of Wolong, the mountains of Tangjiahe are almost gentle, the valleys are broad, and the hillsides slope toward ridges seldom higher than ten thousand feet. The wide vistas had given me a sense of freedom; there not every walk was a scramble. And I had liked the serenity and seclusion. At Wolong, the huge hydroelectric scheme at one end of the reserve (it transmits power to the Chengdu basin but not to Wolong), the villages and heavily cultivated slopes of the Pitiao valley, the large headquarters with its countless staff, the road busy with logging trucks—all rob the reserve of a peaceful atmosphere. By contrast, Tangjiahe is on a dead-end road. Only three hundred villagers lived in the reserve, all near the entrance, and the staff included just twenty people; the area was removed from outside pressure, media attention, and other intrusions. Here, too, we would begin our research afresh, with a new base and a new team. As we sped north I wondered if the dilapidated hut we had selected as research base had been renovated and if log traps for pandas had been built.

Coming with me to Tangjiahe was Wang Xiaoming, a college graduate who was being transferred from Wolong. I did not know him well. During the period of internecine strife among the Wuyipeng staff, he had assumed a protective coloration, submerging himself into his surroundings, doing what was expected of him, no more and no less. Alan Taylor and Qin Zisheng, who would stay a week to establish bamboo study sites, were also with us, as was Xiao Qiu, who for a while would serve as interpreter.

After several hours the road wove between steep-sided hills, their lower slopes covered with coarse grass, brush, and fields, some terraced but others slashed into the hillside and often tilted at seemingly impossible angles. We sped past numerous hamlets, chickens fluttering away and defying death by inches, as we followed the serpentine Baimajiang, the White Horse River, rushing toward the plains. Then the road continued up and down smaller valleys and across low passes, and finally at a village we turned onto a track rutted with irrigation ditches connecting the fields on each side of us. Here the countryside smelled of freshly turned earth and manure. Soon the valley narrowed, its slopes broken by cliffs and stands of trees and with long stretches of river between huts. The river branches at the entrance to the Tangjiahe Reserve, which encompasses about one hundred and seventy square miles of mountains to the north and west. Following the western fork, the Beilu, we reached Maoxiangba, a logging camp until the reserve was established in 1978 and now its headquarters. The drive had taken nine weary hours. Several leaders of Qingchuan county, among them magistrates Sun and Wen, had assembled to welcome us and offer their cooperation, the first of many friendly gestures during the year. Tangjiahe's leader, Yue Zhishun, greeted us with an immense smile, and Teng Qitao, who had been in Wolong some months and would head our team here, was eager to begin our cooperative effort, noting that our research hut and traps were ready.

March 1984. I slept at Maoxiangba and in the morning I went outside shortly after dawn. Here at an elevation of forty-seven hundred feet the temperature was around freezing and the slopes were still brown, with no sign of spring. Someone called to me and pointed across the river. Browsing along the bank was a takin bull. In my many months at Wuyipeng, I had never seen a takin—only rusty wire snares, made of wire much heavier than that used for musk deer, with which villagers had almost eliminated the animals. This takin bull had no concerns, wholly ignoring me on the streambank opposite him, as he bent a sapling over with a sweep of his neck and leisurely nibbled off the evergreen leaves. He had the same ungainly attractiveness as a warthog, except that he stood over four feet high at the shoulder and weighed six hundred and fifty pounds. His straw-colored coat with splotches of gray-black on legs, flanks, back, and rump blended into the dead grass and brush. His coat color indicated that he was a takin of the Sichuan subspecies, different from the dark-brown animals along the slopes of the Himalaya from western China to Bhutan and the golden animals in the Qinling Mountains of Shaanxi Province. However, all subspecies are similar in that they look as if they are constructed from ill-considered body parts of other animals. Assemble the bulky, humped body of a brown bear, the legs of a cow, the broad flat tail of a goat, the knobby horns of a wildebeest, and the black, bulging face of a moose with mumps and you have a takin. Inhabitants of forests on rugged and remote mountains usually above altitudes of four thousand feet, takin have retained their privacy to such an extent that only connoisseurs of the exotic or crossword puzzle addicts have even heard their name. The best general account of the species was still based on H. Wallace's hunting experiences described in *The Big Game of Central and Western China,* published in 1913, so I was particularly interested in becoming acquainted with these animals.

We drove up a narrow track along the boulder-strewn banks of the Beilu. The hillsides looked tattered, covered with patches of brush and saplings and occasional stands of oak, aspen, or other trees. Relatively gentle slopes, especially those with a southern and western exposure, were once cultivated but were now overgrown with grass, weeds, and brambles. In places neat rows of young pine marked attempts at reforestation. From 1965 until 1978, timber was extracted from all accessible areas, and hillsides were either clearcut or selectively logged for conifers. The loggers cultivated their own fields, although groves of tall aspen and other trees typical of secondary forest indicated that agriculturalists had tilled these mountains for decades and probably centuries. As with reforestation elsewhere, the saying "Ten axes and one hoe" applied here too, felling of trees having far outstripped planting. Even when one viewed the hills from the road, it was obvious that previously cultivated slopes lacked bamboo. The brambles might provide fruit for bears and browse for takin, but pandas had been deprived of good habitat. Bamboo would have to be replanted.

About six miles from Maoxiangba, after a sharp turn in the road, the valley widened a little to a place called Baixiongping, the panda plain; *baixiong* or white bear was a widely used name for the panda in China before the 1940s. Crowded between roadside and hill, our new base lay at an elevation of fifty-eight hundred feet. It was a large hut divided into six small rooms plus a larger communal space. Some of the original mud-plastered walls remained, and the rest was a patchwork of boards and reed mats. Each room had a desk, a chair, one or two beds with loose straw serving as mattresses, and a small window. Living here would be comfortable, especially if stoves were installed. Beyond the hut was a shed divided into a bunk room, storage space, kitchen, and the cook's sleeping quarters. As at Wuyipeng, we were in the heart of panda habitat, but here we could roam widely and with ease along the road and into several tributary valleys. It was with pleasant anticipation of the months ahead that I settled into my room. I hung red blankets on the walls to keep out drafts and unpacked my belongings.

Our team was small. Teng Qitao, an energetic man in his thirties, was camp leader. Square-jawed, his hair like straight wire, he was forthright and occasionally fierce, reminding me of a leopard with sheathed claws. Unlike many Chinese, he had never learned to hide frustration beneath a veneer of indifference. Famed locally for his temper, he was jokingly called "uncookable," a play on his name, which means "cook." He would become one of my favorite colleagues. Wang Fulin of the reserve staff, I soon learned, was marvelously adept with his hands, weaving a grass cushion for my chair, bending wire into stove tongs and drying racks, as well as doing many other tasks around camp. Although he was little more than a teenager, Shen Heming, also of the local staff, was soon valued by us all for the cheerful and dependable way he did his work. Hair hanging in his eyes, he

We named this takin bull, who often tarried near our Tangjiahe camp, Lao Pengyou, Old Friend.

During April and May the juicy shoots of umbrella bamboo entice pandas from the ridges into the valleys.

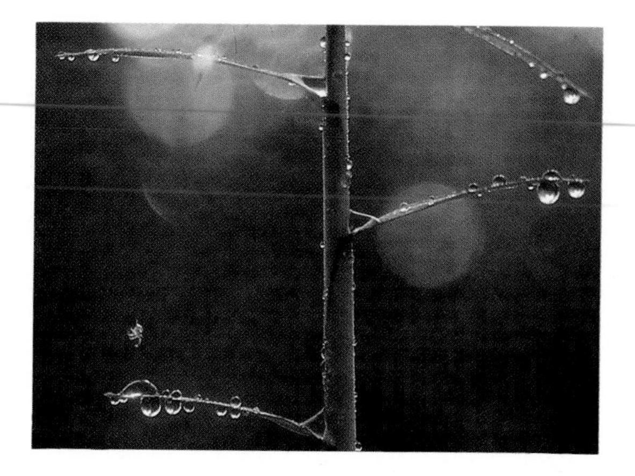

Pandas peel the coarse sheaths of bamboo shoots before eating the tender center. I investigated many such feeding sites.

After the die-off of Wolong's arrow bamboo in 1983, leafless dry stems covered the high slopes.

Many panda young have been needlessly captured and taken to captive breeding stations.

In 1869, Père Armand David was camped near the site of this old Catholic mission building when a local hunter brought him a panda skin, the first ever seen by a Westerner. Since Père David's visit, much panda habitat has been converted to fields.

A sedated panda is wheeled to the operating room for artificial insemination at the Wolong research center.

Staff from the Wolong Reserve transport the body of a female panda I found in a poacher's snare.

Hu Jinchu measures Zhen-Zhen's maternity den in a hollow fir.

Zhen-Zhen investigates our communal hut in search of food.

Zhen-Zhen peers from the window of our research tent.

The endangered golden monkey also shares the panda's habitat.

Zhen-Zhen crouches, head tucked under, while a male, probably Pi-Pi, mates with her.

On a fog-bound slope, Pi-Pi circles a tree to which a female in heat has retreated from his ardent advances.

(following page) To survive in the wild, the panda must have the freedom and security of its mountain forests.

whistled and sang as he checked traps, counted bamboo stems, collected droppings, recorded temperatures, and did many other chores our project demanded. Wang Xiaoming, like most persons posted to Wolong, had vanished into a spiritual black hole, but now, released from that oppressive atmosphere, he quickly revived and surprised me with his initiative and interest.

Within days we were more than just researchers in the same camp; we had become colleagues working toward a common goal, interacting with an easy rapport such as had never developed at Wuyipeng. Partly this was because we lived more like a family, rooming in one house as well as eating together at a table, and partly it was because we had escaped the inhibiting vigilance of a political hierarchy. Although there was one Party member among us, his attitude was relaxed, and consequently there were no dedicated watchers, only workers. When we had visitors, my colleagues immediately became less companionable, more wary. We also had a cook, though if that was his true profession many elsewhere cheered his absence; a local woman replaced him after several weeks, and a succession of other cooks followed.

While Alan Taylor, Qin Zisheng, and Shen Heming set off daily to establish bamboo plots, the rest of us baited and checked the seven log traps. The mountains that March were bleak, mostly shrouded in gray mist. Snow was still a foot deep on the shady side of the slopes. We went up the road, flanked by thickets of willow and *Buddleia,* a tall, brittle shrub that favors logged flats. After about two miles, the road became impassable to vehicles. At that point the main valley forked, becoming the Hongshi and Jiazhihou valleys. The lower reaches of the Hongshi were densely covered with bamboo. Though the valley was heavily logged, good timber remained higher up along the ravines, and it was there we had set our traps. The trails to most traps led up at least one steep pitch, icy, muddy, or both, depending on the season. Clutching saplings and bamboos and finding tenuous footholds while carrying a bag of bait, I always had a sense of athletic accomplishment when I finished my rounds without having slipped.

The dominant bamboo below seventy-two hundred feet resembled Wolong's umbrella bamboo but was somewhat shorter and more spindly. This bamboo, *Fargesia scabrida,* flowered and died in patches between 1972 and the 1980s, with a flowering peak in 1975. The die-off mainly affected bamboo on the lower slopes of some valleys, leaving ridge tops and other valleys, such as the Hongshi, almost untouched. From about seventy-two hundred feet upward grew *Fargesia denudata,* another species of about

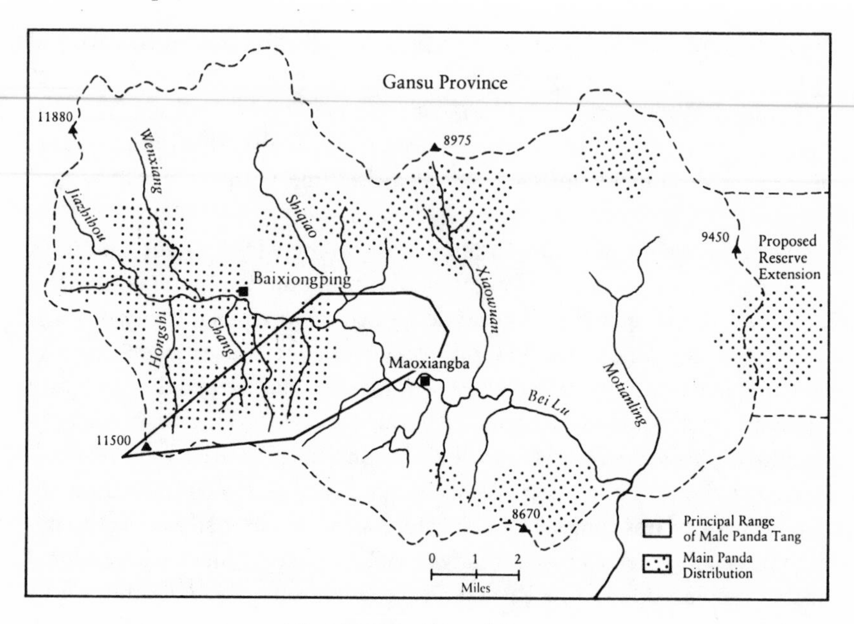

The Tangjiahe Natural Reserve

the same size, though with tiny leaves; it had flowered so extensively in the mid-1970s that only a few patches remained unaffected high on some ridges. Bamboo in the reserve, therefore, existed as a mosaic, ranging in height from tiny seedlings to adult stands of stems six to eight feet high.

Although in mid-March an occasional snow storm still brushed the slopes with white, spring was near. A few primroses were in bloom, and on rare days when clouds vanished, bumblebees zoomed past. Our traps yielded no pandas. I patrolled the road constantly, scanning the slopes in the hope of seeing an animal as it traversed a clearing or sat beneath winter-bare trees. One afternoon, on 19 March, I heard the yips, hoots, and barks of a panda's mating song from several hundred feet above me in a stand of spruce and hemlock. To approach the panda undetected and observe it would be impossible. Mangled by loggers, the lower slopes had grown into a dense tangle of secondary growth, an amorphous mass of brambles with wicked hooks that tore clothes, a labyrinth of branches that raked the face, and vines that not only sewed everything together but also wrapped an intruder as a spider does its prey. Instead, I clambered up a nearby rock slide that had cut a swath through the forest to the road. The panda continued to call as I cautiously moved up among unstable limestone boulders, but suddenly it stopped. Had it seen me? Then I glimpsed the animal, a swaying white rump moving away among the trees.

Just two days later, having returned to base after checking traps, I was told of a panda on a nearby slope. Near our hut an overgrown field swept fifteen hundred feet up the slope; Wang Xiaoming had been there observing a takin bull when a panda strolled by along the forest's edge. I hurried to the meadow and climbed up to where Wang Xiaoming watched the panda eating leaves in a swale among scattered bamboo clumps. Two cars stopped on the road below. It was a Japanese film crew from Nippon Television on a week's visit to Tangjiahe as part of an effort to make a panda documentary. Seeing the sudden swarm of people emerging from the cars, the panda did what it does so well—it vanished. The takin bull stoically ignored the commotion.

Laboratory scientists can schedule work to suit themselves, botanists have no need to stalk or be satisfied with only the spoor of their sedentary subjects, and ornithologists can focus their attention on a nesting bird for hours. Although I occasionally met mammals, Tangjiahe was similar to Wolong in that at the end of a long day of hiking and looking and listening my notebook would be almost devoid of notes, and my pocket might contain just a few droppings of panda, leopard, or wild dog. Admittedly, there were limits to the amount of excitement the contemplations of droppings could provide. Yet there was one exception to this usual tale: the takin.

Takin sign was pervasive. On any hike through the forest, I tended to follow takin trails, for the animals had selected the best routes around cliffs, up hillsides, on ridges. Traveling along, I noted willows stripped of bark and patches of brush torn and broken as if hit by a tornado where takin had fed. Not that takin herds were easy to observe. A strong barnyard odor, or, if I was fortunate, snorts and the thumps of fleeing hooves were usually all that the animals left behind. However, bulls tended to be misanthropic, spending much of the year alone or in twos low on the slopes of the Beilu valley rather than hidden in the forest.

Walking along, eyes to the ground looking for panda tracks, I would sense a presence. Glancing up, I would see a takin bull in dense brush less than twenty feet away, motionless like some tan boulder, peering at me over his curved muzzle, nostrils flared and head held high, obviously prepared to be either benign or displeased. I was grateful that such bulls guarded themselves from spontaneous reaction, giving me time to decide on my own course of action. Usually the animal bolted in crashing retreat; but sometimes he lowered his horns in threat, an emphatic signal deterring any intimacy on my part. Since takin have been known to attack dogs and hunters and might also turn inattentive strollers into casualties, I prudently conceded them the right of way.

One bull in particular was often near our base, placidly foraging there like a domestic cow. He was lean and past his prime, and most of his left horn was broken off. We named him Lao Pengyou, which means "Old Friend." I had first met him on the meadow near camp snuffling among dead weeds in search of the season's first greens. Not wanting to alert him, I had watched from a distance, but since he seemed unconcerned, I moved to within a hundred feet. His placid routine consisted of feeding in the morning and afternoon and resting at midday, when he dozed, chewed cud, and soothed itches by rubbing against trees. When snow covered the ground, he took his rest standing up, his body broadside to any sun, absorbing the warmth of its feeble rays. Forage was plentiful. He clipped branch tips off willow, maple, hazelnut, and other shrubs and trees; gnawed pine bark; nipped leaves off *Buddleia* shrubs and over-wintering sprouts of sage; cropped the coarse tops of horsetails; and, most important, browsed on evergreen leaves of oak, rhododendron, ilex, and a few other woody plants. His size and strength enabled him to reach browse denied to others. Rearing up, he balanced on his hindlegs and cropped leaves eight feet above ground. If a morsel was still out of reach, he used another method: he propped his chest or forelegs against a sapling and leaned forward. Sometimes the stem snapped under such weight, but more often it bent, and the takin simply straddled it, keeping it from springing erect, while he fed. Stems up to five inches in diameter may be broken down in this manner.

While a bull's life conveyed ponderous ease, winter presented a serious food problem. Although they were masters at bush-bashing, takin obviously lost weight during winter, not from lack of food but from lack of nutritious food. To maintain an adequate diet, a herbivore must forage on those plants that at any season offer the most protein and digestible energy for the least amount of effort, and an animal improves its chances of obtaining a high-quality diet if it finds many species palatable. The takin's winter diet was certainly flexible. Digestible energy was not in short supply because takin, like other ruminants but unlike pandas and bears, can turn cellulose into calories. Protein—needed for growth and maintenance—was a problem, however. Evergreen leaves have little protein, just barely enough to maintain the body weight of herbivores. However, evergreen leaves often contain toxic compounds that inhibit digestive processes and prevent some of the available protein from being assimilated by the body. Unless a takin enters winter in good condition, its body may lack the reserves it needs to maintain it until spring, when new growth high in nutrients—double those available in winter—becomes available. Takin did not stuff themselves on nutritious and abundant bamboo leaves during winter. Why not? I have no idea.

Lao Pengyou made it through the winter thin but alive. But that same March we met another bull who was literally a bag of protruding bones encased in hide. He tottered along, feeding on dry sage leaves as if lacking the energy to seek better food. I saw him over a period of days, always farther down-valley as if he was heading toward spring and the life it offered. He was too late. Two weeks after we first sighted him, he stumbled and fell among water-worn boulders at a stream's edge; there, too weak to rise, he died. Poor old fellow. His teeth were worn flat with age, and he weighed a third less than an adult male in good condition.

April 1984. Kay arrived on the first of April, finding not just a shelter but a luxurious hut with electricity. A small hydroelectric plant powered by a nearby creek provided light, though it worked only a few hours a day in those months when the water was high. We also had a small stove, a most temperamental and inefficient one that required constant tending to produce heat. We burned low-grade coal rather than depleting our surroundings of wood.

On reading my journal for that April, I find mostly notes of minor ecological happenings. I find the first spring arrival dates of white-capped redstarts, long-tailed minivets, blackbirds, white wagtails, red-rumped swallows, Peking robins, and other birds. On 7 April, wall creepers flashed their red wings on every cliff, but the following day they left for higher elevations. A week later, the thickets swarmed with leaf warblers (*Phylloscopus*) of several species, small greenish birds that flitted nervously through the undergrowth. There is a note of the date when the drab Elliot's babblers around the hut began to give harsh territorial calls, and of the time when I greeted the first snake of the year, a small *Lycadon* attired in black and white bands. While I listed animals, Kay checked on flowers, the first anemones, clematises, cinquefoils, violets, wood sorrels, viburnums, wild cherries, spiraeas. I made mental notes of where wild strawberries bloomed, in anticipation of a crop not only for ourselves but also perhaps for bears.

By mid-April, the forest shimmered with the soft green of young leaves. Various rhododendrons successively splashed the hillsides with color, pink, red, white, lavender, and purple, and late in the month *Kerria* shrubs were ablaze with yellow blossoms. It was a time of discovery, of exploring a remote part of China where few naturalists had worked. The first to venture into this Sichuan-Gansu border region were the Russian botanist Grigori Potanin and zoologist M. Berezovski. During their expedition from 1892 to 1894, Berezovski obtained several giant panda specimens from the local people, the first records of pandas from the Min Mountains. Reading my daily jottings for April, and for May as well, I realize that this was our most satisfying time on the panda project, the pleasures of nature combined with those of a relaxed and friendly atmosphere. Hu Jinchu and Qin Zisheng seemed aware of this too, and escaped from Wolong as much as possible to work here.

Several of my coworkers were exceptionally diligent about studying English. Most mornings I heard Shen Heming read vocabulary aloud to himself, oblivious to the freezing temperature in his room. Sometimes he came to our door calling "George." When I looked out, he pointed silently to a word in his reader. "Chopstick," I pronounced slowly, and he repeated this once or twice. "*Hao,* good," I said, and with a nod he marched off, perhaps to reappear a while later with another word. The partitions between rooms were of very thin wallboard, and we could hear the others practice too. Teng Qitao and Wang Xiaoming had studied English in college and knew many words, but because they were unused to speaking and hearing English, they hesitated to converse. To help them gain confidence, Kay gave daily "lessons" to one or the other. On sunny days they sat on a bench by our door, chatting slowly in English or reading aloud from a textbook. Teng Qitao and Wang Xiaoming also used English to record their sightings of animals in our camp diary, an amiable way to share experiences. I wished that I had similar willpower and dedication to improve my Chinese. As it was, we communicated in an inelegant mixture of English and Chinese words.

I trained many at Wuyipeng in the basics of field biology but whenever I met former coworkers again they had, with rare exceptions, such as Yong Yange and Qiu Mingjiang, either failed to apply their new knowledge or had become office workers, drivers, or otherwise had switched jobs. I hoped to introduce Wang Xiaoming and Teng Qitao to the pleasures of doing their own small projects, not just collecting panda data as mechanically as they studied English. I saw Wang Xiaoming reading his ecology text over and over, repeating the principles like slogans, instead of going outside to look creatively at something new, to see what there was, not what one was told. I was therefore delighted when he and Teng Qitao initiated a study of takin foods and the nesting schedules of various birds around our base.

After hearing a panda call in mid-March, unusually early in the season, I spent many afternoons walking the road hoping in vain for a repeat performance. Since males sometimes courted after dark, I decided that we needed a tent at the mouth of the Hongshi valley so that we could spend occasional nights there. On 6 April, we all joined our efforts to erect it. There was much noise as we cut poles for the frame, hammered and sawed, and talked loudly. A white spot on the slope above drew my attention—it was the head of a panda monitoring our activity. It was a crisp, sunny day. We lunched by the roadside, the panda eating too, staying in its patch, sometimes visible though more often not. Kay and I spent a number of nights in that tent. From late afternoon to late evening we prowled the valley in cold drizzle or silver moonlight. We heard koklass and tragopan pheasants call, an owl hoot, and a nameless beast crash in a thicket. But the pandas remained silent. We closed our traps for the season without having enticed one.

Our attention now shifted more to takin and bear. Once snow retreated to the highest gullies, I frequently explored several tributaries to the Beilu, especially the Jiazhihou and Wenxiang both of which extended northward to the high ridge that forms the reserve boundary. I enjoyed the intimacy of the valleys. Although long-abandoned fields were tucked even into these remote drainages, the forest remained little altered, and shady stands of spruce, birch, and other species crowded the stream's edge. I followed one side of the stream until I was halted by waters sweeping against a mossy cliff or cutbank and then leaped from boulder to boulder to gain the other side. The gurgling and splashing of the stream hid the sound of my footsteps. Around each bend was a mystery to be solved, and I peered cautiously ahead, hoping to spot bear, takin, serow, or even leopard before the animal sensed danger and bolted. Once I saw a panda munch on two-foot bamboo seedlings that covered the stream bank like meadow grass, but several birds scolded, and the panda, alerted to my presence, fled. I must have stalked around hundreds of bends and surprised fewer than half a dozen animals, yet my anticipation remained undiminished.

Where humus was soft and deep, the woodland flowers grew in great variety, primula, touch-me-not, trillium, mint, buttercup, polygonum, sourdock. There was tall-stalked *Cardiocrinum,* whose large leaves have a copper sheen, and the familiar jack-in-the-pulpit, a peculiar plant that starts its existence as a male and then solves its midlife crisis by continuing as a female. Both the takin and I liked to wander among these flowers. Such young plants were favored takin fodder at this season, and on hands and knees I examined what the animals harvested, which included most plants within reach of their broad mouths. I collected a colorful spring bouquet of the takin's menu to put in my plant press. Whenever Qin Zisheng came to Tangjiahe, she graciously identified my collections, not just the food plants but also flowers whose beauty I admired and wanted to call by name.

One day a driver came from Maoxiangba and in rapid Sichuan dialect gave me a long account, all details of which escaped me except that there was something dead. Since no one was around to interpret, I hopped into his car, thinking an animal had died by the road. After picking up two men at headquarters, we continued a short way downstream, where we began to walk. Fording the Beilu in icy, thigh-deep water, holding our boots and pants overhead, we ascended the Xiaowuan valley, first on a foot trail, and, after forested slopes had hemmed us in, by crossing and recrossing a turbulent stream on swaying logs or leaping wildly. A group of stump-tailed macaques scampered off through the trees, the first time I had seen these monkeys in the reserve. After two hours we reached an overhanging cliff, a primitive shelter where three men squatted by a fire. They had been on patrol, I discovered, and found a dead takin. Unlike Wolong's leaders, those in Tangjiahe—first Yue Zhishun and then Jiang Mingdao—sent out patrols to monitor wildlife and deter poachers. Although I roamed widely, I never found a snare. Once, as Kay and I wandered up a small valley, we came upon three collectors of medicinal herbs, their baskets loaded with roots. They fled, obviously worried about being apprehended for this minor infraction of regulations; in fact, they were so worried that they wrote a signed letter of apology to Yue Zhishun. Decades of hunting had depleted wildlife and even eliminated the red panda, whose fur is still much in demand for hats; but, given protection, species were now on the increase.

We examined the dead takin, a yearling female with spiky horns. She was emaciated, and the lack of fat in her bone marrow was a sign of malnutrition. How could she starve when food plants grew in such abundance? Then I noted that one eye had a cataract; the other had been eaten by a crow. The takin may have been blind.

It was by now too late to return home. While some of us made a bed of evergreen oak boughs beneath the overhang, others started dinner, boiled rice and stir-fried takin pungent with age. As we ate, the head and lower legs of the takin smoked and sizzled in the fire. The hair was being burned off before taking the pieces home to eat. We slept in a row, crowded for warmth. With dawn we rose stiffly to huddle by the fire. Takin for breakfast. Across the creek from us, a yellow-blossomed tree shone against the dark curtain of conifers. I finally reached home at noon, having had a somewhat longer outing than anticipated.

May and June 1984. Shrubs and trees all leafed out in May and herbs were high, hiding wildlife well in the luxuriant forest. As the new growth aged, becoming less nutritious, takin bulls moved up the slopes, following spring to higher elevations. I missed Lao Pengyou's presence. He had seemed to derive a kind of satisfaction from being near our hut. When Kay went to the stream to do laundry, she sometimes labored under his inscrutable gaze; when we used our flimsy outdoor latrine, he occasionally browsed so disconcertingly close that it seemed as if his next step would bring him through a tarpaper wall.

7 May. A memorable takin day. It is one of those rare days when the mountains are at their most lucid. The dark wedge of the peak upstream is crowned by the light of an intensely blue sky, and the leaves of beech on its slopes glow with pale-green fire. I am sitting in the sun transcribing notes when I hear someone, probably Shen Heming, yell that a takin herd is on the meadow high above our hut. We all take turns watching the animals through a spotting scope. As they emerge from the forest's shadow into the open, several young bulls and yearlings become frisky, bucking and bounding; infected by such exuberance, several females leap sedately, though their sturdy bodies, designed for plodding over rough terrain and not for kicking up heels, lack a feminine grace. We know that several small herds with ten to thirty-five animals frequent our part of the reserve. But today more and more takin appear and slowly spread over the meadow until there are about a hundred. With forage so abundant, several herds may have joined. My coworkers soon drifted away to other tasks, and only Kay and I continue to observe the herd at intervals. I settle comfortably into a chair, a cup of tea by my side, utterly content just to watch the animals.

Like males of many species, takin bulls prefer to intimidate an opponent rather than waste energy and risk injury in combat. Two young bulls, playful yet with serious intent, display their prowess. With stiff gait and neck arched far down, one bull cuts broadside in front of the other, moving as if utterly weary, yet displaying his humped profile to full advantage. The other bull is not impressed, he does not turn away to feed or otherwise signal his submission. The first bull tries again, sidling closer. Next the two walk parallel to each other and about six feet apart, but neither is willing to admit defeat. As if on signal, they lock horns, bodies straining as they seesaw back and forth. One breaks away, scrambles uphill, and again meets his opponent horn-to-horn. This time, with the force of gravity added to his attack, he drives the other downhill. Being at a disadvantage, the latter breaks away, and the two race side by side to continue their quest for dominance, whirling, chasing, hooking, pushing.

After observing them awhile, I realize that I have not seen any young. Where are they? The birth season should be over, as young are probably born in April. Finally they come, all together near the tail end of the herd, a tumble of takin, dark brown and fuzzy, quite different in color from the adults. Most gambol after one female and several cluster around another, the youngsters obviously in a kindergarten tended by a baby-sitter while their mothers forage and socialize. I watch until dusk when the herd moves back into the forest. At dawn they briefly come out again, this time all young with one female and so tightly bunched that I cannot count them until she walks on, trailed by sixteen baby takin in single file.

Now in early May we also noted swarms of fleas which lived by the roadside in disconcerting numbers. When Kay first accused me of bringing fleas home from the field, I pooh-poohed the idea. However, a few days later fleas were so abundant that I could observe them land on my pants one after another while I stood on the road. In the past, I had been engulfed by mosquito swarms in Alaska, stalked by hordes of land leeches in Sarawak, covered by an army of minute ticks in Brazil, and played host to battalions of bed bugs in Nepal, but this was a new field experience for me. We now battled to keep the fleas out of our room. When we returned to camp after an outing, we examined ourselves with great care, and, if a bite was found all clothes were dumped out the door. We shook and aired bedding daily, and we developed a nervous tic of constantly scanning our pants. The surreptitious scratching of my colleagues indicated that they also offered homes to vagrant fleas, probably more comfortable homes than I provided because they wore long underwear into the heat of the summer. Fortunately the flea eruption subsided within two weeks.

Mid-April passed, and I still had not seen any sign of bears. Surely with the temperature above 60° F on sunny days and the valleys filled with succulent herbs, the bears ought to be out of hibernation. At Wuyipeng the previous spring I had searched for Han's den tree of the autumn before. I had gone from tree to tree, checking every large one for a hollow base, and in one hollow found a black bear hunched over, its back to the entrance. It was a small bear, semi-awake but still in bed on 12 April. In Tangjiahe, on 30 April, I finally discovered a slushy dropping, and two days later at the mouth of the Hongshi valley, I inadvertently disturbed a bear feeding on the tender new growth of hydrangea shrubs. Bears, it soon became clear, would be difficult to study because they traveled so widely. Animals cruised up and down valleys and over ridges, stopping occasionally to feed but not to linger. Picky eaters, their spring diet consisted of certain herbs and sprouts, at least fifteen species. Once, for example, a bear ambled down a valley for two-thirds of a mile, its broad paws leaving a trail of broken herbs that I could easily follow. The animal ate crunchy stalks of jack-in-the-pulpit and wild celery, and it was particularly fond of the new shoots of a bramble, *Rubus coreanus,* that was common in logged areas. These shoots, which looked somewhat like those of bamboo, were juicy but bristled with red hairs. Sometimes a bear peeled the shoots but more often it ate them whole, which must be like munching on a test-tube brush.

An invitation to the county seat, Qingchuan, to take part in the inauguration of the local chapter of the new China Wildlife Conservation Organization, provided a break in our routine. The national organization was established late in 1983 to promote and raise funds for the protection of wildlife. Hu Jinchu and Xiao Qiu were with us as we left for the two-and-a-half hour drive to Qingchuan. Our route wound through hills among flowering apple trees and terraced fields being planted to rice. In a statistical mood, Wang Xiaoming counted five hundred and eleven curves in the road, not including sudden swerves by Lu Daiming in his attempts to clip chickens and dogs. Drivers in China are pampered and catered to, and none more so than Party members. No one dared fault Lu's driving.

There were many speeches during the inauguration, including my own brief speech on the importance of conservation, which Xiao Qiu translated. Local involvement is essential to conservation, and I hoped that our presence emphasized the importance of the government's initiative. Various groups then presented red scrolls of congratulation to the new organization, the red signifying good luck and happiness, and we lined up for the mandatory group photograph, an important gesture in any communal endeavor. Following was a banquet featuring Qingchuan delicacies: dried fern fronds, noodles made of fern fronds, black fungi. Qingchuan produces more of these thin, rubbery fungi, a favored addition to many Chinese dishes, than any other county in Sichuan. They grow on oak saplings, which are cut in three-foot lengths and placed on the ground.

I was told that a few students wanted to meet Hu Jinchu and me. We were escorted to the school, and as we turned the corner of the building, about a thousand students confronted us, to my consternation, all lined up in neat rows on a sports field. I was asked to lecture on conservation for twenty minutes. The crowd was hushed, less in anticipation of what I had to say than to hear me talk in English. I was the first live foreigner they had seen or heard. Though unprepared for a lengthy speech, I did my best to make them aware of the need to treasure their natural heritage. I encouraged them to help the new wildlife organization, respect and protect living things, plant trees, and be proud of the pandas and other wildlife in their county. After Hu Jinchu and I completed our talks, girls tied the triangular red bandanna of Young Pioneers around our necks.

One evening not long after our visit to Qingchuan, Wang Fulin bicycled from Maoxiangba to inform us that a panda had been found dead far up the other main valley, the Motianling, an area I had not yet visited. That valley, like ours, had a road, but, after loggers left, it was not maintained and bridges soon collapsed. The following morning, we drove as far as we could and then walked, the sun hot on our backs. The panda was in a patch of bamboo where it died without a sign of any struggle at least two months earlier. Only hair, pieces of skin, and the skeleton remained. We bundled up the bones and walked back to wait for our driver, who had gone on to town for supplies and had not yet returned.

Clinging to the slope above us were several huts; Teng Qitao knew one of the residents, an old bear hunter, who invited us to wait for the car in his home. His hut was surrounded by fields, most of them tiny plots molded to the uneven terrain and planted in potato, cabbage, lettuce, maize, wheat, barley, onions, peas, and beans. The walls of his hut were of vertical boards, the roof covered with tarpaper, the floor of dirt. On each side of the door, shaded by walnut trees, were stone corrals with pigs. At one end of the hut, on a cleanly swept area, grain was being dried on grass mats. There were several bee hives. Caged beneath an upturned basket sat a mother hen, the basket raised just enough to allow her downy chicks, but not the hen, to wander in and out. Two bony dogs, so different from the sleek pigs, skulked around the corner as if ready to dodge a stone or foot. The soot-blackened main room had a fire pit in the floor; above it hung a kettle suspended from a hook. We sat down on stools covered with bear and musk deer hide. A poster on one wall depicted a soft, fairy-like girl riding sidesaddle on a tiger that symbolizes male power, and on another wall such miscellaneous items as a feather duster, a curved knife, and a long-stemmed pipe with a brass bowl were stuck in cracks between the boards. Through an open door I could see into a bedroom, dark except for a shaft of light from a tiny paneless window. The *kang* at one end was piled with quilts, and there was a red chest, decorated with two swans, on which stood an alarm clock and transistor radio. A picture of Zhou Enlai was on the wall beside a framed collage of photographs. When I asked our host about his family, he showed me the photographs—his mother, his eight children, himself at various ages, and four Chinese actresses. Only his wife, a son, a daughter-in-law, and a grandchild lived with him now.

He was a hospitable and voluble old man, his face as brown and wrinkled as the walnuts he served us. A druid image, he had a wisp of beard and only four yellow canines remained of his teeth. He served us hot water sweetened with honey. His wife, dressed in dark blue homespun and sneakers, brought a plate of pickled onions. He then rummaged in a basket and drew forth a plastic bottle filled with liquor, *bai jiu,* which we declined.

"I am self-sufficient," he noted with obvious pleasure. "I grow almost everything I need except for salt and cooking oil. And for that and also rice, I trade some of my produce."

His life was hard, his surplus meager, yet he appeared satisfied with what he had, accepting his condition and fate; by keeping his needs and desires modest he had found contentment. I envied him that. He and his wife were prepared for the future: outside in a shed were two massive coffins, each one consisting of just four thick slabs and two end pieces. I had been told that within a year or two all villagers would be removed from the reserve, the families scattered and absorbed into new communities. Did he know that he might soon be uprooted? I hoped that he could die here in the security and peace of his old home, yet I also admired the government for its efforts to rehabilitate the reserve. A conservation issue is seldom clear-cut, it often involves moral ambiguity, and the happy old man on his sunny hillside still dwells uneasily in my mind.

The social event of that season was on 26 May. Qingchuan's leaders held a banquet for forty people at Maoxiangba, a festive occasion full of generosity and goodwill that affected me deeply for its was held in honor of my birthday. First we gathered for a group photo, and two small girls in coral dresses gave Kay and me each a bouquet of plastic flowers. Then we adjourned to a meeting room where laudatory comments about cooperation and hard work were made, and Hu Jinchu, who had come from Chengdu, expressed a wish for many more years of association. I was given a marvelous assortment of presents, an overwhelming bounty from everyone, the most I had ever received on a birthday. Among the presents was a box of medicinal roots said to have special powers to cure the headaches of intellectuals. I was intrigued. But since no one there knew either the dosage or proper way of eating the root, I never used it; at any rate, I seldom have a headache, perhaps a reflection on my intellect. Four Western-style cakes thick with white frosting had been brought from Qingchuan. A total of fifty-one burning candles encircled two of these cakes. In trying to blow both out with one puff, I ran out of air halfway around the second cake, much to my chagrin. The two little girls danced and sang a local song to great applause, and Kay led everyone in "Happy birthday." During the banquet I had to toast everyone individually, always emptying my small glass. As everyone insistently reminded me, tradition demanded that I drink fifty-one glasses, one for each year of my life. It was a memorable occasion, or so I have been told.

Halfway between Maoxiangba and our base, at the mouth of the Shiqiao valley, were several patches of *Bashania* bamboo, a species with large leaves. New shoots appeared in mid-April. On the first of June, Hu Jinchu and Wang Xiaoming found that a panda had recently eaten shoots there—tall shoots already hard at the bottom—and two days later someone again saw a panda. Hu Jinchu and I were in neighboring Pingwu county on a panda survey when, on 8 June, we received word that a panda had been caught in Tangjiahe. We returned immediately, reaching Maoxiangba in the afternoon. There we learned that Teng Qitao had again seen the panda at the mouth of the Shiqiao valley. He and twenty-eight villagers encircled the animal in a bamboo patch, and it conveniently retreated into a cave where it remained. Teng Qitao and Wang Xiaoming met us on the road. To prevent the panda's escape they had spent a restless night at the mouth of the cave, lying in a mosquito-infested hollow over which they had spread a tarp to keep out rain. I peered into the cave, really a rock cleft in the hillside that was no more than five feet deep, low and moist, with roots pushing through the ceiling. Wedged into a corner sat the panda, staring back at me with unblinking eyes. I suggested that a carrying cage be brought from Maoxiangba while I fetched the immobilization equipment.

The panda remained mute when I injected it. Five minutes later I wormed into the cleft and tied a rope around one hind foot; slowly we pulled the animal out and placed it inside the cage, where it sat groggily, almost filling the space. It was a middle-aged animal with yellow somewhat blunted teeth; Teng Qitao said it was a male. Two persons reached through the bars and held his head upright while I fastened the radio collar. "This went well," I said to myself with relief, and wondered if we should release the animal now. It would be dark by the time he had fully recovered from the drug. Best wait until morning.

Among those who crowded around was a man I did not recognize; he was a surgeon from a hospital in Qingchuan, who had been summoned the day before because the panda had thrown up several roundworms. I was now informed that the panda would have to be wormed and then held at least a week for observation.

"But all pandas have worms," I replied with obvious consternation. "It doesn't mean they're sick. We can't remove a panda from the wild just because it has a few parasites."

Unfamiliar with animals, the surgeon had played it safe in his recommendations and, following his lead, Hu Jinchu and Teng Qitao acquiesced. I agreed only to hold the panda captive overnight. We took his cage to our shed and gave him rice porridge, bamboo, and water. He sat hunched and silent.

By morning he had not eaten in spite of having fasted for nearly two days. When Kay poured a little water on his muzzle he at least licked at that. He gave occasional pathetic grunts. We gathered before the shed to discuss his fate.

"He is obviously ill. He hasn't eaten," said Teng Qitao.

"Of course he hasn't eaten," I countered. "He is scared. Before we captured him he obviously ate well."

"We can't let him go in this condition," persisted Teng Qitao.

"I appreciate your concern; I know you are worried about the animals's health," I said with a combination of suppressed anger and apprehension. "But if you keep him in captivity, he will become ill. And then you can't return him to the wild soon. The project at Wuyipeng has been seriously damaged because of misguided decisions by leaders to take three pandas into captivity. One of them died. I wasn't there then to argue with the leaders, but I'm here now. If you take this animal into captivity, there is no point in continuing our work in Tangjiahe."

With this ultimatum, I turned away. Kay was nearly in tears at the thought of what might happen to the panda. Since I had taken responsibility, the others conferred for less than a minute before deciding that the panda would be given his freedom. We immediately loaded him into the car and took him to the place of capture. As soon as he saw his familiar bamboo patch he became agitated, his dejected look gone as he reared up and looked around. When we opened the door, he rushed out.

We checked on him that afternoon and again the next day. He had remained nearby. We had visitors from Hong Kong, the first foreign tourists to enter Tangjiahe—Mary Ketterer and Ken and Sue Searle, who had also joined us in the Jiuzhaigou Reserve, as well as the well-known British conservationist Ian Grimwood and his friend John Brocklehurst. I was pleased that they could hear the beeps from Tangjiahe's first radio collared panda. Teng Qitao, Wang Xiaoming, and I quietly entered the bamboo to make certain the panda was feeding. Fresh droppings littered the thicket.

We named the male Tang-Tang—for the reserve, for his sweet temperament (the word also means "sugar"), and for the Chinese dynasty best known for its treasures.

Having sat in his bamboo patch for a while, Tang suddenly began to look for greener pastures. He crossed the Beilu, angled up a ridge and vanished into a maze of mountains, his route taking him east past Maoxiangba. We picked up a weak signal for a few days, and after that he vanished. We clambered up several ravines to radio locate him, but we picked up only leeches. Tang's behavior was unlike that of the sedentary pandas in Wolong. He might teach us something new about pandas, assuming we could find him again.

To learn Tang's whereabouts would be Ken Johnson's first priority, for Kay and I were due to leave Tangjiahe on 23 June, she to return home and I to begin wildlife work on the Tibetan Plateau with a two-month survey. During my absence, Ken would move from Wolong to Tangjiahe. Ken had been using the project's foot snares in Wolong to capture pandas, and now he would also try his skills on Tangjiahe bears.

It often rained at night in late May and June, but the days were sunny with enervating heat and humidity. Kay and I usually took a stroll toward evening when the air was scented with honeysuckle, our steps leading us often to the mouth of the Wenxiang valley. There, shaded by pine and hazelnut, was a huge boulder. One side of its surface was deeply carpeted with moss, covered at this season by hundreds of lavender *Pleione* orchids. We often tarried there. A meadow nearby had a wonderful crop of wild strawberries, both white and red. Picking the berries was a sporting venture. One did not know when plucking fingers might encounter one of the numerous *Trimeresurus* vipers that also favored the strawberry patch. Irascible creatures, they lay curled along the mane of grass in the center of the road and struck at our boots as we passed. At this season, too, dogwood was in flower, the trees so white with blossoms they were like snowdrifts at the forest's edge.

During the night of 19 June it rained heavily. Lying in bed, we could hear the river's murmur turn into solid noise, a roar as if from a never-ending train. It rained all day, another night, and the day after. The river became a brown cataract, a rage of water, every creek a churning torrent, every seepage a waterfall. Across from our hut was an old landslip overgrown with weeds, and we watched as sections of slope slipped away piece by piece, reopening an ancient wound that would take decades to heal. The serene beauty of the river had become an energizing force as it stormed down the valley. Exhilarated, Kay and I stood in the rain to watch the violent waters splash our feet and listen to rolling boulders rumble like muted thunder in the depth. The rain continued through a third night. Toward morning the hut trembled when a nearby section of river terrace crumpled. One massive landslide or a washed-out bridge might close the road for days. We packed hurriedly and left for Maoxiangba a day earlier than planned, and then continued to Chengdu.

September and October 1984. In mid-September we once more drove toward Tangjiahe. Lao Chen, our new driver, a tall, quiet man with a head gaunt as if carved from wood, was at the wheel; he was a "steady" driver, the Forest Bureau had assured me. Along the Beilu River the first yellow leaves announced autumn. Yet tall, purple anemones, *Anemone hupehensis,* bloomed incongruously along the wayside, like mementos of spring. Huge piles of driftwood lined the roadside, the logs scoured by rock and sand until they were pale as bone. On 3 August it had rained five inches, raising the river even above the level reached during the June flood; indeed the high-water mark on the bridge was more than thirteen feet above normal.

Glancing through the team's record book of animal sightings, I noted that summer was not a good time to observe wildlife. Anyway I was primarily interested in news about our two radio collared animals, a panda and a bear. Neither could now be found, I was informed. We had lost Tang's signal in June, but Ken had briefly located the animal again in July in the hills to the south, outside the reserve. Tang's range was at least nine square miles in size, a large area compared to that of Wolong's animals. He certainly did not need to travel so widely to find food. Could Tang be a subadult who had not yet settled into a home? I had taken Teng Qitao's word that he was an adult male. If he was a subadult, then I should replace the collar with a larger one, giving his neck room to grow. In retrospect, the animal had seemed unusually small for an adult. The more I thought about the matter, the more uncertain I was whether Tang was adult or subadult, male or female. Teng Qitao succeeded in replacing Tang's collar the following June. He was indeed an adult male, a small one weighing just one hundred and forty-nine pounds.

Our second radio collared animal was Kui-Kui, named after Li Kui, a choleric and powerful character in the ancient Chinese novel *Shui Fu Zhuan*. He was a black bear, a big male, and by all accounts most appropriately named. He made such an impact on my colleagues that they ceased trapping as soon as Ken left, afraid to handle a snared bear. Kui chewed off the tree to which the snare was attached—a tree six inches in diameter. He lunged and roared and fought the air between himself and his tormentors. Ken, who had handled many American black bears, was suitably impressed by the force of his violence.

They immobilized Kui at night. The drug took effect more slowly than anticipated and several doses were necessary before the bear reclined. Still alert, Kui swiped at Teng Qitao, hooking his shirt with a claw and drawing him in toward his jaws. Coming to the rescue, Hu Jinchu and Xiao Qiu pulled Tang Qitao in the other direction—and won. Later, as Ken straddled the dozing bear to attach the radio collar, Teng Qitao took a flash photo. This promptly awakened Kui, who grabbed one of Ken's legs with his forepaws but abruptly fell asleep without biting. Finally, back on his feet, Kui roared and charged, scattering everyone in the darkness, and in a final spasm of ursine umbrage mauled Ken's fallen cap. In his report, Ken noted laconically that Kui "had satisfactorily recovered."

Disgruntled, Kui continued his wandering. Signals revealed that he was active mainly in daytime, and settled down for a night's rest at around 9 P.M. He remained in the nearby hills just long enough to be monitored for three days. Like Tang, he then roamed widely out of the reserve to the south, his range encompassing at least twelve square miles of rough terrain.

Shortly after my return, I walked along the road, antenna held aloft as I probed every valley for a signal. I quickly found Tang in the same bamboo patch where we had caught him in June, and located Kui on a ridge crest in a stand of oak. Kui soon disappeared again, whereas Tang remained, feasting on bamboo leaves. On one occasion, I found Tang asleep in a pine close to the road. When he heard the click of my camera, he gave me an intent look, but then, quite unconcerned, dozed off again, much as Zhen sometimes behaved in Wuyipeng.

Coming home in the fog one afternoon, I almost collided with a takin bull. He snorted and stepped back. His left horn was broken off: it was Lao Pengyou. Having deserted our valley in May, he probably spent the summer high on the slopes where he may have given up his solitary ways to join females for the rut. Peak rut is said to be in August, a time when swirling rain clouds engulf the mountains. But now Lao Pengyou was back for the winter.

Alan Taylor joined us for a week, bringing his wife, Kristin, and his guitar. During the day, Alan and Shen Heming checked bamboo shoots in the study plots—pandas had eaten sixteen percent of the shoots since they appeared in July. And on some evenings Alan and Kristin sang *Blowing in the Wind, Clementine,* and other songs we particularly enjoyed. After my summer's absence, I seemed to be getting into a routine again. To gain those insights upon which a study depends, to feel the rhythm of seasons, and to note the slight changes upon which animals base their lives, one must be constantly in the field, not coming and going like a migrant bird. Yet in late September I had to leave Tangjiahe once again.

I flew to Switzerland to discuss the future of the panda project at WWF headquarters. With almost four years of field research completed, it was necessary to evaluate our efforts and establish guidelines for future work. Research so far had focused on the natural history of the panda to give us a baseline of information about the species. We needed to continue this work because many unanswered questions remained. But it was now time to shift the project's emphasis to conservation, to the preparation of a master plan for the conservation and management of the panda and its habitat. Such a plan would have to take into account various human factors—the ecological, social, and economic influences on the panda both inside and outside of reserves. After all, the panda's home was now surrounded by a dense human population whose impact on the forests was great and would only increase.

We decided on several goals. Pandas would be censused throughout their range, the distribution and abundance of each bamboo species would be mapped, and conservation problems and possible solutions would be delineated for each area. Surveys to achieve these goals might require two to three years. These and other suggestions were to be presented to the Chinese for discussion.

That done, I met Kay in Germany and together we traveled back to Tangjiahe, arriving there on 11 October to spend a final three months with the pandas.

31 October. A routine workday. At dawn, close to seven in the morning, I step from the hut to glance at the sky. Low clouds on the slopes, new snow down to the seven-thousand-foot level. The thermometer registers 35° F. I shiver involuntarily at the sight of snow, yet look forward to winter.

Although the valley is lovely now, with leaves of autumn swirling red and orange and gold, snow will concentrate wildlife low on the mountains. While Kay starts a fire, I walk a short way up the Chang valley that projects south of camp. I have set two foot snares there to capture bear, though I do not really expect to catch one; bears are at lower elevations at this season, picking acorns. But I enjoy a purposeful morning walk. The thickets bordering the old logging trail are hushed, my steps silent on sodden leaves. A gray-headed bullfinch perches on a willow, his feathers fluffed, and an Elliot's babbler skulks away. Both will spend the winter with us. A black-faced warbler scurries among brambles. He, like the many Kessler's thrushes, finch-billed bulbuls, Peking robins, and other summer birds that now restlessly crowd the Beilu valley, will soon be gone. Breakfast is ready when I return, our usual rice porridge, roasted peanuts, hard-boiled eggs, steamed bread, and pickled vegetables.

Teng Qitao has posted a weekly duty sheet on the wall. I am to check three log traps this morning and the more distant bear snares in late afternoon. The trip up the Hongshi valley will take about four hours, and I leave immediately with a new short-term assistant, Lu Xiaolin. Our team is breaking apart. Wang Fulin has left, Wang Xiaoming will return to college within a month, and I will leave the project in January. These anticipated changes naturally create an unsettled mood, but I also sense less commitment to the future of our program and this worries me. Work at Tangjiahe has barely started. The most important discoveries often come not after a year or two of study but after five or ten years, when subtle and rare events in an animal's life may provide crucial insights. When I express my concerns for the future to Teng Qitao, he replies reassuringly, "I will be here, if not as a leader then as a worker." The traps are discouragingly empty again. The panda population in Tangjiahe is small, perhaps only fifty to sixty animals. Unlike their habit at Wuyipeng, the pandas here travel on ill-defined routes, making it difficult to anticipate their movements and place traps accordingly.

A bowl of noodles for lunch. Afterward, Wang Xiaoming and I walk three miles down the road. As usual we scan the hillsides for takin. A solitary bull rests at the base of a cliff, his coat almost indistinguishable from the autumn foliage. No herds are visible, though they sometimes come into the valley at this season to lick and gnaw soil at certain spots. Soil may provide minerals or somehow help detoxify any noxious plants a takin has eaten. Perhaps for similar reasons, takin have consumed most of an abandoned house by the roadside, chewing holes into the mud walls until they collapsed. Fortunately our hut is mostly of wood, or Lao Pengyou might demolish it. He is often nearby, closer than ever, in spite of noise, stoves belching smoke, and persons wandering around. Kay, who remains uncertain about his temper, once commented as she watched him from our window, "This is the way I like to observe takin, sitting comfortably by the fire drinking tea."

On our walk we also see the white flash of a tufted deer's rump patch, usually all these deer reveal of themselves as they bound out of sight into a thicket. Small and solitary, tufted deer are fairly common in Tangjiahe, but we learn little of their secretive habits. They are certainly one of the most attractive of all deer, gray-brown and adorned with white markings on ears and legs. Males have ebony canines protruding from beneath their upper lip and tiny horns hidden in the tuft of hair on their crown.

One lone oak looks as if a dozen huge birds have built leafy nests in it. A few days earlier, we had observed a black bear in this tree. Squatting in a fork, the bear had broken and pulled branches toward itself with a forepaw, and plucked any acorns with its mouth. Then it peeled the acorns with its teeth, spitting out the shells and eating the kernels. Discarded branches were pushed into the fork beneath its feet, creating a nest-like platform. On seeing the bear, we had hoped it was Kui, with whom we had lost contact over a month ago, but it was not. Carefully we check the tree again. I am uncertain why we do so, for the bear has long since moved on to other oaks in distant valleys. Almost every oak has broken branches, some trees so damaged that only trunk and branch stumps remain. In their autumn fervor to harvest the nutrient-rich acorns, the bears have reduced their food supply of future years. Not being conservationists, their only concern is to lay on fat quickly before their winter's sleep.

We leap from boulder to boulder to cross the Beilu River, and scrabble up to a grove of trees in which we have concealed a bear snare beneath leaves at the base of an oak. The bait—a chunk of goat tied to the trunk behind the snare—remains intact. Only a weasel has extracted a small tax of meat in payment for our intrusion into its domain. Farther downstream, two other snares are empty too. Our day's field work completed, Wang Xiaoming and I now trudge back up the road toward home.

Tang remained in his usual bamboo patches at the mouth of Shiqiao valley. His behavior puzzled me. Our work, and continued monitoring the following year by Teng Qitao and Don Reid, showed that he resided in that half-square-mile area except from June to August, when he moved to higher elevations. Yet to our knowledge he neither sought females nor was visited by them during the mating season. An area must have suitable den trees before a female can settle down in it, and Tang's domain, like most slopes in Tangjiahe, had been logged. During my roaming in the reserve I found not one den tree near bamboo; one female had chosen a rock overhand to shelter her newborn, a drafty, unprotected site. Tang's activity schedule was also somewhat different from that of other pandas. Near his favored haunts was the shell of a building, one room of which was occupied by a gnome-like old man with a toothless smile who grew cabbages and tended an orchard of apples and pears. We used an empty room of this hut from which to monitor Tang. He spent more time at rest than any panda we had studied, and like Kui he slept much of the night. Once when Kay and I arrived to take over the monitoring, Teng Qitao and Wang Xiaoming were worried because Tang had been inactive for the past thirteen hours. Was he ill or dead? Much to our relief, after an eighteen-hour nap he became busy again. Tang moved so little—he remained all November in a two-acre patch—that he certainly expended little energy. Kay and I observed him on the far bank of the Beilu on one occasion, only his head and his arms visible. Bending an elbow was his most vigorous activity as he pulled stems close with one forepaw and pushed leaves into his mouth with the other. For an hour we watched his lantern-bright head, until he suddenly stopped feeding in mid-chew. The wind had shifted and he obviously scented us. Quietly he stepped out of sight. Tang was a reminder that animals may not only vary in their behavior from one area to another but also are individuals with their own idiosyncrasies.

November 1984 to January 1985. On 5 November we caught a bear in a snare above Maoxiangba. It was a small bear, but the ruff framing his face made him look large, and his fury was most certainly outsized. Never had I encountered such an angry animal. With screaming roars he charged Jiang Mingdao and me again and again, his attacks abruptly thwarted by the end of the snare cable. In frustration he bit large slivers of wood from the tree. When at last he was quiet for a moment, I shot a syringe into his thigh and three minutes later the animal was asleep. He was a male about three years old and weighed one hundred and fifty-five pounds. The snare had caught not his wrist but only three toes. Had he pulled free during an attack, the consequences could have been interesting. Since a bear gains and loses so much weight in the course of a year—it loses at least a fourth of its body weight during hibernation—the neck consequently expands and contracts in size too, making it difficult to fit a collar to the animal. I fastened the collar, hoping the bear would not pull it off, and moved far back. Within minutes, he was on his feet, woofed at us once, and hurriedly departed. We called him Chong-Chong, another temperamental character from the same novel that provided Kui's name.

November was our bear month, mostly devoted to Chong. By mid-month most leaves were off the trees, and each snowfall encroached lower on the valley. Still Chong roamed, feeding on acorns, butternuts, hazelnuts, and whatever fruit he could find, not yet showing an interest in hibernating. Perhaps he had already selected a rock cleft in one of the many cliffs around Maoxiangba. Finally, during the last days of November, Chong ceased to travel, though his faint signal from high in the hills remained active. However, on 8 December the signals became inactive and remained that way; he was asleep.

A bear's adaptations for conserving energy during hibernation are extraordinary. The animals do not eat, drink, defecate, or urinate. Heart rate decreases, body temperature drops slightly, metabolism is reduced by a quarter. Combustion of stored fat furnishes energy for metabolism and provides metabolic water for maintaining appropriate levels of blood and other fluids. In spite of all this, a bear does not sleep deeply; females remain alert enough to give birth and care for offspring.

Teng Qitao and I decided to locate Chong's den. We climbed up a steep-sided valley, scrambling for several hours up a creek bed broken by waterfalls glazed with ice. High on a ridge, wind-whipped in a snowstorm, we listened to Chong's erratic signal, sometimes clear and at other times faint, as if at a great distance. It was too late that day to make a search for the den. We descended by another slope that was deep in winter's grip, lifeless but for the footprints of a panda, the animal alone in these heights as if abandoned. The panda's dependence on bamboo forced it into a life-style very different from Chong's, the bear now cozily tucked away in his den until spring.

The diets of Tangjiahe's pandas and bears differed almost completely. Pandas subsist on bamboo, bears on herbs, fruits, and nuts. A panda might occasionally sample wild celery, a favored bear food, and bears ate bamboo shoots, but that was the extent of overlap in their food habits. Bamboo represents a constant food source in unlimited amounts; herbs and nuts grow scattered and are seasonal. Analyses showed that bear foods contain, on the average, two to three times more available nutrients than bamboo. Given similar digestive abilities, these bears thus obtain two to three times more energy per pound of food eaten than do pandas. Acorns are particularly nutritious, but a bear must work for this food: it must find, pick, and shell four to twelve thousand acorns, depending on their size, to get a daily ration of eleven pounds. It has been estimated that an American black bear, which is ecologically similar in food habits and activity to its Asiatic cousin, may in one day eat enough calories to provide sufficient energy for five days of hibernation.

Pandas generally have to spend more than twelve hours a day eating but, because food is so abundant, they need not travel far. By contrast, scarce and patchy food resources require bears to maintain large ranges and travel widely in search of ripe berries and isolated oaks. Yet their food is so high in quality that bears can store fat and hibernate with only eleven hours of activity or less a day, the free hours allowing them a good night's sleep. High-quality food enables bears to reproduce at a younger age than pandas, as young as three and a half years if good food is plentiful. They average larger litters, and a female often raises two or three offspring.

Bears have remained opportunistic and adaptable, geared to a boom-or-bust economy, evolving a life-style that has enabled them to settle and thrive in many different habitats. The panda has become a specialist, it has chosen security over uncertainty. But by so doing it has lost its need to explore, to be observant, to try something new; it has tied itself to a fate without a horizon. Both bear and panda are triumphs of evolution, but in this age of environmental destruction, it is the adaptable species that has a better chance of survival.

On 14 December we awoke to a transformed world, the brown slopes and skeletal trees obliterated by the season's first heavy snow. I walked up the valley to check traps, my steps as muffled as if I were walking on a quilt. A spotted forktail screeched and fluttered away in a flash of black and white as I passed a rivulet. Later, headed toward home, I saw the tracks of someone running down the road. Shen Heming had been checking the other traps; perhaps one had caught something. Soon Kay came toward me, calling happily.

"There's a panda in Trap Seven!" The trap was on the crest of a ridge traversed by a least two pandas. After months of effort, we had at last caught one.

As I removed a log from the trap to allow room for maneuvering the jabstick, the panda roared and a black paw flashed through the gap. Cautiously I peered in and faced two eyes peering out. Finally the animal, a small one, turned aside, and I injected the drug.

"*Nü*, female," said Teng Qitao as we pulled her out. Her nipples were long and dark. "She's had babies," I noted. "Are you sure?" queried Kay. "She's very small."

She weighed just under a hundred and fifty pounds, about a quarter less than the females at Wolong. Teng Qitao had told me that pandas in this region are much smaller than in Wolong, and, judging by Tang and this female, he was right. While the others returned to camp, Teng Qitao and I waited for her to recover from the drug.

"We name the panda Xue-Xue," said Teng Qitao. *Xue* (pronounced Shu-eh) means "snow," a good name to commemorate this day.

With two pandas and two bears radio collared, Tangjiahe had a solid project, a good basis upon which to continue and expand the work. As if reading my mind, Teng Qitao said somewhat frantically, "You stay here. In January and February we catch three pandas, four pandas," Soon all responsibilities would be his, and he was concerned.

"You know how to do all the work very well," I replied. "I'm not needed anymore. Besides, Alan and Ken will return to Tangjiahe next spring, and another foreigner will collaborate with you too."

Finally Xue poked her nose out of the trap, then ambled uphill and out of sight.

For most of December the weather was dismal with leaden low clouds. Around noon a pale sun sometimes reached the hut but soon slipped behind the ridge, leaving us for another day in shadow, icy and mute. The hydroelectric plant stopped functioning when the creek froze, and without the light our rooms were gloomy. I spent as much time outdoors as possible, looking for tracks in fresh snow and hoping to observe some animal.

The road was a remarkably busy highway for wildlife, though most creatures used it mainly at night. I found the tiny, paired tracks of rats and mice trickling from one side to the other, the meandering furrows of shrews, the bounding imprints of weasel, the dainty hoofprints of tufted deer, and the bovine prints of takin. A rare leopard or golden cat demarcated the road's edge with its tracks. And occasionally small packs of wild dogs trotted purposefully along, stopping now and then to sniff some spot or communally defecate. I observed where a panda had padded across the road and where a yellow-throated marten, its splay-toed tracks distinctive, had investigated a brush pile. I had met martens in pairs by the river in late October and early November, undulating along like miniature otters. Why had they chosen the riverbank for what appeared to be their mating season? Once, as I paused, there was a sudden raucous and piercing "Waaaii" close to me; I leaped forward and was ready to take evasive action when a little grizzled hog badger scuttled away. Marten, palm civet, hog badger, golden cat, and other small carnivores in the area interested me. They were worth a study, but it would remain for someone else to do.

Early in December, before the ground froze, Kay and I had carefully dug up a small pine to be used first as a Christmas tree and then planted in front of the hut, a living memento of our transitory presence. As in 1982 at Wuyipeng, we decorated the tree with family ornaments. On Christmas Eve we had a small private feast—a bottle of wine and bread and sausage we had brought from Europe in October. The following day there was a banquet in the communal room. Hu Jinchu and Qin Zisheng came from Wolong to join us. The Christmas tree with presents beneath it sparkled in the light of many candles. We had a quiet dinner, a family dinner, with steamed carp, smoked tofu, dumpling soup, and numerous other dishes. There were private toasts but no official speeches, a banquet both happy with friendship and sad with farewell.

I parted from my colleagues here with reluctance and regrets. Hu Jinchu and I had worked together amicably for four years, our shared hardships transcending cultural boundaries. Qin Zisheng had made many of our trips more cheerful by her buoyant presence, and Teng Qitao and Shen Heming, each in his own way, had helped to make our months in Tangjiahe a pleasure and a success. The candles burned low and the dishes were empty; outside snow swirled. The door suddenly opened and in from the storm came Jiang Mingdao and Yue Zishun, heads and shoulders capped with snow, after a walk of six miles to join our festivities. Both had done much to make our year here a happy one, and Kay and I toasted them in special gratitude.

As our day of departure approached, each observation and each action had a special poignancy. Placing the last dried panda dropping on the tray for sorting, I offered it to Kay, but she allowed me the honor. A tent had been erected in the Hongshi valley from which we could monitor Xue, who for the past three weeks had remained near where we caught her. On 6 January, Kay and I went to spend the night in that tent, our last night of listening to a panda's monotonous routine. To our delight, we met Lao Pengyou near the tent, as if on a social call two miles beyond his usual thickets. He moved stiffly and his hindquarters were lean. "Poor thing," said Kay. "I'm afraid he won't make it through this winter."

The slopes looked blue-gray in the evening light. With the temperature in the tent at 20° F, we ate our *mantou* and sipped cocoa while tucked into our sleeping bags. It was Kay's birthday. "I've even taken you out for dinner on your birthday." I noted.

On 8 January, I checked my last trap. The next day, before loading the car, I walked down the road one final time. Golden monkeys whined on the slope above me and I spotted three of them in the bare crowns of birch. The tracks of a panda led me to several bamboo clumps where the animal had fed, biting off more than fifty stems to browse on the leaves.

Toward noon we drove to Maoxiangba, and the following morning we continued down along the Beilu and out of the reserve. It seemed as if some final gesture was needed to mark the end of a phase of our lives, but we simply drove on, our own work done but the project not yet over.

A Buddhist inscription on a temple in Chengdu reads: "Things under the sun are never finished and are finished as unfinished."

12

Crisis

1983 to 1985

Our research at Wolong and Tangjiahe was interrupted by a crisis, that of a major bamboo die-off.

We had been on a survey of various panda habitats in the spring of 1983, when I noticed that in Baoxing county, southwest of Wolong, the arrow bamboo had begun to mass-flower, a periodic event that occurs only every forty-five years or so. After flowering, the bamboo would seed and then die within a few months. Since arrow bamboo is a preferred food of pandas, the death of this bamboo could cause widespread starvation among the animals. In the mid-1970s many pandas starved after bamboo flowered in the Min Mountains. Another tragic loss had to be prevented at all costs. I suspected then that the die-off was not confined to Baoxing county, and indeed, on my return to Wolong, I found the bamboo in flower there too. The forest was brown with inflorescences as if covered by tall autumn grass. Three-fourths of the arrow bamboo was dying. However, green patches remained, enough to support the pandas for months. I had no need to worry about the immediate future of Wolong's animals. Even if the remaining arrow bamboo should flower the next year, pandas would still have umbrella bamboo available at low elevations. But umbrella bamboo could also flower. To monitor the situation, two teams, in addition to the one at Wuyipeng, were stationed in remote valleys.

We were seriously concerned about pandas elsewhere in these mountains. Little was known about the extent of flowering in other parts of the panda's range, though we assumed that arrow bamboo throughout the Qionglai Mountains had been affected. We had to prepare for a major effort to rescue starving pandas.

With admirable speed the Ministry of Forestry, under the leadership of

Vice Minister Dong Zhiyong, initiated several actions. Preparations were made to construct refuge camps, holding stations at which starving pandas could be rehabilitated for release back into the wild. Every county with pandas established a committee, led by the deputy magistrate, to coordinate rescue work. Teams of four to six persons each began to roam the hills to locate starving pandas and rescue them. A number of communes, eager to be of assistance, also formed their own teams. Any villager who found and led a rescue team to a starving panda would receive a reward equivalent to about one hundred dollars, a considerable sum. Local people were prohibited from cutting bamboo in panda habitat, and there was widespread publicity on behalf of saving pandas.

The arrow bamboo would not be dead until late autumn, and Hu Jinchu and I planned to check on conditions in various parts of the panda's range during the winter. Preparations made, we could only wait. I returned to the United States for a three-month break.

News in the press that summer of 1983 reported additional emergency plans. The Chinese government had allotted funds to feed starving pandas by placing food, such as grass and meat, in suitable localities. And special teams would be sent afield to rescue baby pandas and raise them in captivity. On reading of these plans, I wondered how enough food to maintain pandas could be distributed daily in rugged and remote mountains. The scheme to capture baby pandas troubled me. My Chinese colleagues and I had different perceptions of captivity for animals. The rigors of nature hold no attraction for most Chinese, who generally adhere to the Confucian belief in man's ability to improve on nature. Several had told me that pandas would be better off exchanging the hardships of the forest for captivity; after all, the animals would then be happy living securely behind walls with a roof over their heads, good food, no enemies, no cares. This is a logical concept in a society that built the Great Wall. But I believe in the freedom of creatures to live unfettered, with whatever tribulations that might entail.

It was November 1983, time for me to return to China.

Before proceeding to Wolong, I attended a meeting in Beijing between WWF and Chinese representatives to discuss cooperation during the coming year. The Chinese revealed plans to build ten holding stations for starving pandas, and they noted the need for one hundred vehicles to assist with rescue efforts.

"We have not been involved in preparation of the emergency plan," commented Charles de Haes. "It would be useful if George could see for himself what is needed."

The suggestion was not welcomed by the Chinese; they do not confuse material assistance from a foreigner with accepting guidance.

"We mainly rely on our own efforts to rescue pandas," replied Li Guilin, a tall, austere man and Wang Menghu's superior. This was the first time he had participated in a project meeting. "WWF should, however, share certain responsibility in the rescue work." He suggested that WWF donate twenty pickup trucks and half the rescue costs.

"We have been urging for three years that an emergency plan be prepared. Nothing was done," said Charles de Haes. "There is no way I can agree to the request. But it is not impossible to get twenty vehicles. I promise you I will do my best."

It was clear that WWF would not be asked to participate in the actual planning for the emergency or in the rescue efforts. In the seesaw of enmity and alliance between WWF and China, the balance at that time was tipped toward enmity.

In an issue of *Time* magazine, dated 28 November 1983, I had read a piece about the Wolong pandas. The animals were said to be suffering from indigestion because they had started to eat ordinary grasses. Two had already died. And "Chinese scientists, aided by World Wildlife Fund, are undertaking emergency measures. One tactic: leaving roasted pork chops and goat meat on the mountain slopes in hopes that the pandas will turn from their normal vegetarian diet."

As soon as I returned to Wolong, I had a meeting with Hu Jinchu and others to discuss the crisis, and I also checked on the accuracy of the *Time* story. Yes, Qin Zisheng had found grass in droppings but only a few at one small site. No, pandas in Wolong were still fine, none had indigestion and none had died. Yes, more than a ton of pig bones, sheep heads, and other offal had been scattered to lure pandas down into umbrella bamboo. I felt pandas would no doubt descend on their own when they were ready, as they had done in the past, and that few would find such bones. But weasels, martens, and other small creatures would delight in the bonus of meat scraps.

Since there were plans to build holding stations, I had asked William Conway to design a small and temporary one. In addition, Emil Dolensek had prepared a list of emergency procedures for treating a debilitated panda, such as giving it sugar water for quick nourishment. I handed the material to the Chinese leaders. One told me, "Once we have money from the government, we can make plans for constructing holding stations." It was clear that cheap facilities were not wanted; I had been careless in my assumption.

I returned easily to the winter routine at Wuyipeng. There was Tian Zhixiang's welcome shout calling me to a meal: "*Chefan,* Georgie." There were the evenings crowded around the communal fire. Daylight hours were spent hiking the trails in arrow bamboo, now a desolation of dead stems, brown, brittle, leafless. I roamed widely to map patches of bamboo that had not yet flowered and found a large area on the far side of the Zhuanjing valley. Our radio collared pandas had concentrated there.

It snowed on 21 December, the first heavy storm of the season, and the temperature dropped to 15° F. The forest was muted under its blanket of snow as I tracked Zhen hoping to see the small pawprints of an infant where she had bedded. But there were none.

Then it was Christmas, lonely this year without Kay, but the camp staff made it a festive occasion. They draped the hut with colored streamers. At one end stood a small spruce decorated with paper flowers and candles. Lai Binghui came, having made the steep climb up from headquarters with two cakes baked in Chengdu. The banquet was as usual loud and long, featuring many courses and speeches and toasts. That the staff with warmth and generosity had prepared a Christmas just for me touched my heart. The Chinese have a special ability to move you with unexpected kindness.

The day after Christmas, Shi Junyi and I left on foot to check bamboo conditions around Niutoushan, Cowhead Mountain. Carrying our sleeping bags, we headed southeast across the Fangzipeng plateau, with its closed ranks of dead bamboo stems, and then up toward the ridge bordering the Choushuigou drainage. The ridge was swathed in cloud. No warmth and little light penetrated the dull fog that engulfed us. Winds whipped the snow, and we were soon encrusted with it. For several hours we plodded along the crest, stabbed on by the wind, uncertain where we were, knowing only that somewhere below, in the Yayadi valley, was a hut where a panda rescue team lived.

Toward dusk we scrambled downhill, hoping to find some sign of the team. Shadows closed in as we entered a thicket; soon it would be dark. But with luck we crossed a trail; following the fresh footprints, we came to the hut squatting almost invisibly in the snow at the edge of a clearing. The four team members, led by Shu Jiegong, whom I already knew, greeted us cordially. They insisted that we take their warm spots by the stove and they made us tea, which we sipped gratefully. The hut was simple, so spartan most bears would have disdained it as a winter home. The walls were of grass mats and the roof of tarpaper. Half of the room was taken up by a *kang* or sleeping platform of poles, the other half mostly by the stove and firewood. A pile of frozen Chinese cabbage occupied one corner; a slab of pork hung from the rafters. I admired the good humor, endurance, and patience of our hosts under such conditions. For the night we all slept on the *kang,* arranged like sardines in a can; beneath the *kang* several chickens clucked softly as they settled to sleep.

A rooster startled me awake by crowing cheerfully within a foot of my ear. Wind-driven snow pelted the hut, and when I peered out the door flap there was only a blur of gray. After a breakfast of rice porridge, we all climbed toward the ridge. This forest had been logged and only scrub remained, bowed with snow, all contours soft and round. Fog smothered us. I was shown two log traps, unchecked for days. The team had been ordered by Hu Jinchu to catch pandas and move them needlessly downhill to areas with more bamboo. Unable to see much in this weather, we returned to camp in the early afternoon. There the hours passed slowly. Gusts blew snow into the room. The Chinese smoked, chatted, and drank tea, and several tutored me in Mandarin. In the evening, before seeking the warmth of my sleeping bag, I checked the weather; snowflakes danced like fireflies in the beam of my flashlight.

It was still snowing the next day. We hiked into a neighboring valley where we followed a streambed, rough with ice-covered boulders, silent with frozen waterfalls. A few patches of umbrella bamboo grew on the hillsides, and higher up was a little unflowered arrow bamboo. We found no fresh panda spoor; animals had possibly moved into another valley to the northeast.

I also visited Hanfunding, the base of another rescue team, to check on bamboo conditions. Travel was laborious in that valley with a foot of snow on the ground and undergrowth almost impenetrable, as stems laden with snow leaned into each other. A panda's trail crossed ours, a deep furrow in the snow that vanished into a bower of bamboo. There was still much umbrella and arrow bamboo in the area; a rescue team was not required here.

Although these surveys were brief, they showed that pandas still had unflowered arrow bamboo available. More important, there were other bamboo species at low elevations to which pandas could turn as an alternative food. I believed that Wolong's pandas would survive the bamboo die-off well.

However, newspapers gave a different impression of conditions in Wolong during 1983 and 1984.

> Chinese wildlife workers have captured an elderly and starving giant panda and put him in a panda farm to gain weight, Xinhua News Agency said.
>
> The panda estimated to be 18 years old and named Huahua, was lured into a cage baited with roast beef and mutton as he climbed a mountainside in Wolong Nature Reserve, searching for food.
>
> He will stay in the Hetaoping farm, with other old and weak pandas, until the end of the current famine caused by the premature flowering of the arrow bamboo. . . ." (*Hongkong Standard*)

Hua was actually one of our study animals, taken in full health without cause from an area with ample bamboo.

> "The pandas are in danger of literally starving to death here," Lai Binghui, director of the 770-square-mile Wolong Nature Reserve, said the other day. . . .
>
> "There is very little for them to eat, and some are too weak to find it," Lai said. "About 60% of Wolong's pandas are on the verge of starvation. As they grow even weaker, they are also likely to become victims for predators or simply fall off cliffs. . . ." (*Los Angeles Times*)

When I left the panda project early in 1985, two years after the onset of flowering, the remaining arrow bamboo in Wolong had not yet died, contrary to my expectations. Rescue teams had not found a single starving panda; villagers had not reported any weak animals to claim a reward. The research team at Wuyipeng monitored conditions closely and found no evidence that the pandas had problems.

How did the pandas cope with the sudden disappearance of more than three-quarters of their favored food, and with their bamboo supply reduced to isolated patches? Ken Johnson and his Chinese colleagues found the answer, and it was mundane. The pandas continued much as before. They remained in arrow bamboo where, as usual, they ate mostly old stems in spring, leaves in summer, and old shoots and leaves in winter. Their daily activity cycle continued unchanged. They did not even move greater distances; however, instead of meandering through the bamboo, they travelled directly from patch to patch. Wei, Pi, Long, and Zhen retained their previous ranges, although they focused their activity in certain new areas, naturally those with unflowered bamboo. Once Pi made a seven-day excursion out of his range into a valley to the east as though checking things out. Obviously enough arrow bamboo remained to meet the pandas' nutritional needs, so much so that they largely ignored umbrella bamboo except for shoots in spring.

But one female behaved differently. She was a subadult named Li-Li who had been collared in April 1984, a year after the onset of flowering. A month later she had moved nine miles to the northeast, coming back to the Wuyipeng area for brief periods only twice that year. We could only speculate that, like Pi, she perhaps made occasional excursions from her distant home near Niutoushan, or that, being subadult, she had not yet settled down.

Hua-Hua was also atypical in his behavior. After his release, he spent much time in umbrella bamboo, his food preference different from that of the other pandas. While in captivity, he had been fed mostly umbrella bamboo and may have learned to like it best.

A panda faced with a shortage of arrow bamboo had three choices: expand its range, emigrate, or shift its diet to another bamboo. The remaining patches around Wuyipeng did not flower in the years immediately after the main die-off (and still had not done so by the end of 1991), but pandas may have been cropping this scarce resource to the limit even though there were fewer animals in 1985 than in 1981, because poaching and natural mortality had decimated the population. Not until the winters of 1985–86 and 1986–87, three to four years after the die-off, as Don Reid would note, did the pandas change their food habits. They ate more from each arrow bamboo stem, making better use of what was available, and they descended to lower elevations to feed on umbrella bamboo in winter, something they had seldom done before. Wolong's pandas were fortunate that in a time of crisis most could stay home and merely switch to an alternative food.

In 1987, the arrow bamboo seedlings were four to five inches tall, a low green carpet. They were still too small to serve as panda food, especially when hidden by snow in winter. By the late 1990s, they will once again cover the slopes in tall stands, offering pandas both food and seclusion.

But in some areas outside Wolong, the pandas were not so lucky.

I had been in Baoxing county when arrow bamboo began to flower in the spring of 1983, and the following winter I had permission to return for a week. Clouds were low and there was a sprinkle of snow on the slopes when on 2 January 1984, Hu Jinchu, two other Chinese, and I returned to Baoxing. Officials briefed us on the crisis. As yet no pandas had died, but everyone was worried. With an estimated two hundred pandas, the county was a major stronghold of the species. About ninety percent of the four hundred square miles of arrow bamboo was dead, and other bamboo species covered only about thirty square miles, mostly near settlements. Several pandas had been reported near villages eating such atypical foods as grass, maize stalks, and a leather jacket.

Yaoshi, the timber camp at which we had also stayed the previous spring, was bone-chillingly cold, the rooms unheated except for a small iron pan with glowing charcoal. We lined up at a kitchen window for a bowl of rice topped with stir-fried vegetables. The icy communal dining hall discouraged lingering; it was warmer outdoors. But there a loudspeaker blared with mind-numbing noise for hours on end.

Much of the arrow bamboo in the drainage had died, but the other bamboo at low elevations remained abundant. The pandas should have no problems here. To the west, however, in the Xi drainage, pandas were said to be in trouble.

The two drainages are separated by a long day's walk over the crest of a high range. With a local guide in the lead to show us the trail as far as the divide, we plodded upward through a foot and a half of snow. The mountains were lost in cloud. The lower slopes were logged, but higher up we walked through forest, the massive firs and the trunks of rhododendrons spectral in fog. At ten thousand five hundred feet we stepped abruptly out of cloud into the blinding radiance of a clear sky. To the north and east lay a corrugation of ridges, grass slopes, and gray rock, almost snow-free in the blaze of sun. Towering above them, Sigunian's peak cut the sky like an ice adze. The trees became ever more stunted and gave way to grassland. Finally, we reached the divide at twelve thousand feet. Below us a river of cloud hid the Xi valley; beyond, a jagged range blocked the horizon. The Chinese took off their wet sneakers to dry them in the sun, and barefoot they wandered through the winter grass and white everlasting flowers. We tarried, none of us wanting to descend into the clouds, but finally we left.

At first the trail plunged into a gorge deep with snow and broken by landslips; then the pitch became more gradual and a broad path used by loggers led us down the mountain. A panda track crossed our route. Soon forest gave way to fields. Below, in the gathering dusk, huddled the hamlet of Zhung Gang. Unlike the stone houses of Tibetans on the other side of the ridge, the buildings here were of wood, the walls of vertical boards as in Wolong. As we entered the hamlet, we passed a shop selling rope, clothing, aluminum pots, paper, candy, and cigarettes—the usual items. Shaggy ponies stood tethered by a gate. There was a hostel. Inside, we settled ourselves on stools around a fire pit in the floor, a layer of smoke hovering just above our heads. People crowded into the room to look at us and discover the purpose of our visit. An old man wearing a black turban and a coat of black homespun squatted down next to me and gently grasped one of my wrists as he gazed into my face, his mouth open in a wide smile. He continued to beam, his hand still on my wrist, saying nothing. Then someone nudged him, mumbling something, and he then explained that he was happy to see me, that he had not seen a foreigner since he was a young man. He was now seventy-four. In 1929, he remembered, the Roosevelt expedition came to this village and camped just upriver but was unable to find a panda and soon left.

We had a bowl of noodles and cornbread for dinner, as fox-faced dogs darted around our legs seeking crumbs. Afterward I was taken into a storeroom filled with baskets of dry medicinal roots that had a pleasantly spicy aroma. In a corner was a bed that someone had generously vacated for my use. Outside the room was a latrine that opened directly into a pigpen below.

The morning was crisp, 24° F, with the valley in shadow and peaks in sun as we hiked upriver. Sometimes brush-covered slopes lined the streambanks, but at intervals there were broad river terraces and alluvial fans at the mouth of tributaries. Fields had been scraped into stony soil. Many homes had verandas along which maize drooped in golden clusters. After about two hours we left fields behind when the valley narrowed. Low-elevation bamboo had occurred this far only in rare patches, and now it too ceased, giving way to arrow bamboo, most of which was dead.

The scarcity of bamboo in this area worried me. Unless pandas climbed over the crest into the Dong drainage, something few would probably do, how could they find enough to eat? Of those far from a village, few would be rescued. Along the upper Xi river, with only one bamboo species available, pandas were at the limit of their existence. In the past, a bamboo die-off probably had caused starvation—and now it might do so again.

Traveling down the Xi valley, we were at first in the shady chill of a canyon that was flanked by precipitous slopes broken by cliffs and scraggly islands of brush and trees. The valley then broadened and its gentle hillsides were cultivated, the soft green of winter wheat imparting a springtime aura to the drab terrain. Logging and agriculture had not only destroyed all panda habitat on the lower slopes but had also eliminated travel corridors of bamboo that would have permitted pandas to move in seclusion from one ridge to another. Here, as elsewhere, inexorable population growth was threatening the future of pandas and of all wildlife.

In the months ahead, a number of pandas in Baoxing county were to die of starvation and several were found weak from hunger. Retained briefly in captivity until they had regained strength, many were then released into areas with bamboo. Had these pandas been radio collared, they would have provided valuable information about their movements in new terrain, essential knowledge for successful reintroductions. For unexplained reasons, radio collaring was not encouraged, and now no one knows the fate of the released animals.

The plight of the endearing pandas starving in remote, snowbound mountains captured the world's attention and imagination. Saving the panda was a readily comprehensible problem, unambiguous, specific, somehow manageable in a time of wars and environmental disasters. China's news releases on all aspects of the panda crisis received world coverage.

Headline after headline emphasized the pandas' plight.

SAVING THE PANDAS . . . SHORTAGE OF BAMBOO THREATENS GIANT PANDAS . . . RESCUE EFFORT UNDERWAY TO SAVE GIANT PANDAS . . . PANDA IN PERIL . . . EMERGENCY MEASURES TO SAVE GIANT PANDAS . . . FOOD RUSHED TO STARVING GIANT PANDAS . . . EMERGENCY FUNDS FOR SAVING PANDAS . . . MORE MONEY NEEDED TO HELP IN PROGRAMME FOR SAVING GIANT PANDAS.

Many news stories emphasized body counts of dead and rescued pandas.

—So far, 14 sick or starving giant pandas have been saved in the drive.
—The official news agency Xinhua and newspaper Guangming Daily quoted Vice Minister of Forestry Dong Zhiyong as saying 21 pandas were found dead in the wild and six others died after being found. He warned that the famine was getting worse, Xinhua said. The rescue operations will last at least 10 years.
—Conservationists have reported success in saving 80 giant pandas.
—Forty-two giant pandas have died in a famine that began last year.

And a few news items stressed heart-warming concern for pandas by local people, as this one from the *Hong Kong Standard* on 14 December 1984:

> A giant panda visited the mountain hut of Li Huaxian, a lacquer tree grower, ate the host's rice as an uninvited guest and left for the wilderness in great contentment.
>
> This is a frequent scene in Baoxing county, Sichuan province, home of the giant panda, one of the world's rarest large mammals.
>
> Zhan Jiliang, a hunter of Qiaoji township, has hung up his gun and sent away his three hunting dogs. He was glad to do this for the panda, though it meant a loss of more than 1,000 yuan a year in his income.

Catastrophes followed by appeals for help, whether they be floods in Bangladesh or droughts in Ethiopia, arouse public compassion. The drumbeat of publicity brought in many donations both from within China and from other countries. *Newsday* wrote on 17 June 1984:

> The government has committed itself to the long-term and expensive goal of preserving the panda as a species. Last year, the government allocated $410,000 for the panda rescue, and the Sichuan, Shanxi [sic] and Gansu provincial governments spent another $100,000 in local funds. This year, the central government has committed $1 million for operating expenses and an equal amount for construction of panda feeding and breeding stations. Another $100,000 is to come this year from the Sichuan provincial government.
>
> The central government has committed itself to maintaining this level of expenditures for at least the next two years.

Individuals and organizations within China also donated generously. Railroad workers collected the equivalent of twenty-three thousand dollars among their two million members. Businesses using the panda as a trademark on such products as candy, flashlight batteries, thermos bottles, toilet paper, flit guns, and thermometers collected money. Calligraphers and artists donated their works. Half a million Shanghai school children had a "panda donation day." The magazine *China Reconstructs* quoted a letter from kindergarten children in the city of Tangshan. "The money enclosed was given us by our Moms and Dads as New Years gifts, but we are giving it to the giant pandas so that they can have beautiful kindergartens to live in."

Overseas help soon arrived too. Ocean Park, an aquarium and amusement park in Hong Kong, gave about seventy-five thousand Hong Kong dollars to the panda rescue campaign. WWF-Japan donated twenty pickup trucks, and the Japanese government provided approximately two hundred thirty thousand dollars. The *San Francisco Chronicle* reported, "At a White House ceremony Tuesday, Nancy Reagan accepted a check for $9589.93 representing donations from American school children to the "Pennies for Pandas" program. The First Lady will hand the check to Chinese officials in Beijing when she travels to China with the president April 26."

Early in 1984, during the height of this media blitz, I was first in Wolong, then in Baoxing, and finally in Tangjiahe where we were establishing our new research program. I became uneasy about the relentless publicity of a crisis that I considered local. The starvation of even one panda is, of course, a tragedy. However, it would harm rather than benefit the pandas if widespread rescue operations included areas where the animals were not in danger. The *China Daily* reported on 16 December 1983:

> Gong, deputy head of Sichuan Province's Forestry Bureau, said that more sick and starving giant pandas have been found recently.
>
> In late November, he said, a dead panda was found in Tangjiahe of Qingchuan County. It was only six years old, weighing 40 kilograms. This is the fifth victim of starvation found in China since October, according to Gong who is also deputy head of the province's panda rescue group.
>
> The first death was discovered in Pingwu County of Sichuan in early October. . . .

Contrary to this news item, I knew that there was no bamboo shortage in Tangjiahe, that no panda had starved. And I had been told that the cause of death of the panda found in Pingwu was not known. Apparently any dead panda was being listed as a starvation death. With rescue teams in the hills and villagers reporting panda remains to officials for the first time, a body count would be high—and there would be no way of knowing how many of the total had actually starved.

I was, in fact, perplexed about the proclaimed crisis in Pingwu county. Arrow bamboo is not a principal species there, and other species had flowered mainly in the mid-1970s and had since regenerated. Local bamboo die-offs continued, however, as we had seen the previous year in the Jiuzhaigou. A number of Chinese reporters visited us in Tangjiahe and all talked about the panda crisis in Pingwu. But none had gone into the field; they had merely interviewed officials in town. A year after the onset of the emergency, no biologist had as yet evaluated actual conditions. I had long wanted to do so, but it was not until June 1984 that Hu Jinchu and I received permission for a week-long survey there.

We drove from Tangjiahe to Pingwu, a two-and-a-half hour trip. The county leader held a banquet in honor of our arrival. Since our time was limited, I suggested that we visit only those areas that had been most seriously affected by the bamboo die-off. It was agreed that Zhong Zhaomin, a forest official much involved in panda rescue work, would show us several such places.

3 June. We labor up a ridge about halfway between town and Tangjiahe. It is humid and the sun is fierce. The slopes have been logged, leaving a tatter of brush and scattered trees. But there are small thickets of *Indocalamus* bamboo, and in one we find fresh panda droppings. Higher up is another bamboo species, almost all of it dead, the dry stems still standing almost a decade after flowering. There are few seedlings. At Wuyipeng, Alan Taylor had discovered that bamboo seeds need the moist coolness of a forest canopy to germinate. Exposed to the force of the sun, logged hillsides become too dry and seeds die. Therefore logging, even though it may not eliminate bamboo directly, may ultimately deprive pandas of their food supply when the bamboo fails to regenerate well after a die-off. We reach the crest of the ridge and follow it, grateful for the cool there beneath a canopy of beech and rhododendron. There is much bamboo, both unflowered stands and patches of seedlings a foot or more tall.

In a hollow is a lone dove tree, fifty feet tall, ablaze in white. The tree's flower is insignificant, but two large, white bracts up to six inches long and three inches wide embrace it. Agitated by a breeze, the many bracts look like a flock of doves taking wing. Ernest Wilson, who in 1899 had reached China with a commission to obtain the seeds of this tree, likened the bracts to "huge Butterflies hovering amongst the trees," and he found the species "at once the most interesting and beautiful of all trees of the north-temperate flora." The elation of seeing one of these rare trees remains with me as we make our weary descent into the valley, where at a stream we cool our sweaty faces.

That evening I conclude my notes with the following concern: "They say a panda was found starved on the hill we surveyed and a young rescued on the opposite slope. I don't believe it. No panda could starve with so much bamboo. The one died of other causes and the young was taken for the reward. A worrisome thing."

4 June. Bamboo is patchy and scarce on the hills and pandas, if any, are rare. A Severtzov's grouse flushes from beside our trail; it reminds me of the ruffed grouse in our Connecticut woods. In a shallow depression are four eggs, large as a chicken's and colored a pale tan with occasional speckles.

Our car waits for us at a village. It is the Dragon Boat Festival, and local leaders invite us to eat the traditional *zongzi,* glutinous rice wrapped in bamboo sheaths. Villagers crowd around. "See, he's just like any other man," says an oldster to a child beside him. "Except for his big nose."

5 June. In rain we drive up along the Baima River on the road that leads to the Wanglang Reserve. Halfway to the reserve, we halt to hike up the Shangyanwo valley. Its lower parts have been heavily logged but, further in, where the valley narrows to the width of the stream, forest remains. We climb a steep ridge to the crest at ten thousand feet. Most of the bamboo is dead, the tall stems draped with yellow-brown moss, and the ground, too, is covered with moss. Noiselessly we walk through the forest, the trees eerie in dense fog. A carpet of bamboo seedlings, only two to three inches tall, grows among the dead stems. Our enthusiasm damped by heavy rain, we give up on this day's survey and tramp back down the mountain.

Today, as on previous days, I have occasionally found a clump of bamboo in flower, out of synchrony with the dead stems all around. Chinese accounts describing the bamboo die-off in the Min Mountains in recent years had claimed that only one species of bamboo, umbrella bamboo as at Wolong, was supposedly involved. Yet this species does not occur at Tangjiahe, and I note that instead two other species had flowered. Now, examining and comparing flowers, I feel certain that not one but at least three kinds of bamboo had burst into bloom concurrently, a dramatic mass flowering that helped to explain the extent of the panda die-off in the mid-1970s.

6 June. In the Wanglang Reserve we check on areas we have not visited previously. From a crest high above the valley, we have a wide view of the mountains. Clouds have balled around the summits of the limestone peaks to the northwest, and forested ridges extend far to the northeast into Gansu. Viewing the terrain from these heights, I can appreciate the raw power of the 1974 earthquake that shook this region. Slopes carry wounds of reddish earth as if slices have been neatly carved off. Below, valleys have been partially dammed by a chaos of boulders, soil, and splintered trees. Cliffs wear jagged scars of a lighter gray than the surrounding rock where huge chunks have seemingly been bitten or wrenched off. Even the placid pandas must have been nervous during this upheaval. A little more bamboo has flowered since the major die-off a decade ago, but to my relief, Wanglang has no renewed problems.

I wrote a report to WWF summarizing what we had heard and seen during our survey: "In contrast to the extensive flowering in the Qionglai Mountains, Pingwu County had no major bamboo die-off in 1983. There was flowering on some ridges above 2900 m in 1982, perhaps causing local food shortages, but little else. Rather than facing a new crisis, pandas are still involved with problems created by the die-off in the mid-1970s, though in many places new seedlings, 50 to 100 cm tall after a decade of growth, provide pandas with food again."

Ken Johnson, who surveyed in this region extensively in 1986, came to a similar conclusion in his report, noting that "bamboo regeneration is excellent in most areas and should provide the base for expansion of these panda populations in the future."

This visit to Pingwu reinforced my impression that there was a lack of coordination in the rescue program. China had made an outstanding effort to mobilize its resources. However, Beijing's leaders mainly sent directives to and had meetings in Chengdu. Leaders there sent directives to counties where officials delegated details to those of lower rank who had little understanding of what was needed or wanted, or even of how to proceed. Actions therefore tended to be based on scant information; lacking detailed instructions, each county went its own way. Convinced by the publicity that all pandas were in dire straits, counties had rescue programs where none were needed. Besides, with so much money suddenly available, all counties wanted a share, crisis or not. As Wang Menghu commented about the difficulties in coordinating efforts from Beijing: "Distant water cannot put out a nearby fire."

To my dismay, a large permanent installation for holding captive pandas was built at Tangjiahe in 1984, even though the region had no bamboo problems. Pingwu constructed two small facilities. And a major one was built in the Baishuijiang Reserve in Gansu, just north of Tangjiahe at a cost of over one million dollars. The *China Daily* noted that "the farm has eight air-conditioned feeding rooms, eight playrooms, nurseries for baby pandas and a kitchen. . . A veterinary hospital at the farm has an x-ray room, laboratory, isolation ward and pharmacy." Four captured panda cubs had already been moved to the installation.

The construction of the Wolong research center, rather than coordinating efforts on behalf of the panda—the purpose for which it was built—had stimulated other counties and reserves to emulate rather than collaborate. What an utter waste of scarce conservation dollars. And of pandas.

The rescue operation, begun with such good intentions, had become a lament. The Chinese realized this and by late 1984 were trying to correct mistakes, as Associated Press correspondent Mort Rosenblum reported in the *Dallas Morning News* of 23 September 1984:

> The last wild pandas, crowded into shrinking mountain top islands and menaced by poachers' deadly snares, face threats far more serious than the publicized bamboo famine, foreign and Chinese experts say.
>
> "The worst problem is a lack of seriousness and knowledge among some of our researchers," said Wang Menghu, director of reserves in the forestry ministry. "But we are determined to overcome this."
>
> Authorities now say they overestimated the bamboo famine, but it is still a problem.

Wolong remained a major conservation problem. Although the bamboo die-off had not much affected the pandas, human activities in the reserve continued to have a serious impact. There were at that time two communes in the reserve with a total of six hundred and forty-seven households comprising some three thousand seven hundred people. The steep-sided valleys provide so little agricultural land that the government supplies annually many tons of grain to supplement local harvests. The villagers use the forests extensively: cutting timber, collecting medicinal herbs, killing wildlife for meat, musk, and skins. The panda's habitat is shrinking, and many pandas die in poachers' snares.

Robert de Wulf from Ghent University in Belgium compared satellite photos taken of Wolong in 1975 and 1983. In those eight years, fourteen square miles of forest had been destroyed. No one knows how many pandas died in snares. Clearly the species has an uncertain future, even in China's best known and largest panda reserve.

The Ministry of Forestry developed a plan to protect the forests: resettle the two hundred and eighty-two households from the upper to the lower part of the reserve. As a first and immediate step, the one hundred households closest to panda habitat, a total of five hundred and ninety people, would be moved and their abandoned fields would be replanted with trees and bamboo. The people could not, I was told, be moved entirely out of the reserve because no alternative place existed.

To entice the people into resettling, the government would construct new brick-and-cement houses for them; provide both a primary and a secondary school; build a hydroelectric station to provide power for heating, cooking, and lighting, decreasing the need for wood; and create more pasture for increased livestock production. In effect, about ten percent of the reserve would be destroyed in exchange for improving conditions in critical panda areas. Since the new area had limited potential for agriculture, and some was already in use, it was not clear how all the new settlers could make a living.

China enlisted the aid of the World Food Programme, which agreed to provide tons of rice and vegetable oil for a total value of seven hundred and seventy thousand dollars. The food would be used as payment in kind for the resettled families and the laborers who were to be brought in to construct homes and plant trees on denuded slopes. In addition, the bamboo die-off presented a perfect opportunity to restore forests degraded by logging, as Alan Taylor has pointed out. Dense bamboo prevents trees from regenerating. But now, with so much arrow bamboo dead, planted saplings would have a chance to grow. The World Food Programme's plan was feasible, though no one knew if the villagers could be moved. The government finds it more difficult to deal with minority peoples than with Han Chinese. Certainly the plan was not enthusiastically endorsed by all, as noted by Jeff Sommer in a *Newsday* story of 17 June 1984:

> "It's no good for the pandas to have people living so near," said Lai Binghui, director of the reserve. . . .
>
> When pressed, the government can be as tough as any in the world. That it so far has not chosen to take a hard line is evident in the defiant tone of many of the Wolong peasants, who do not accept the government edict.
>
> "We're not moving," said Li Tiaxing, 35. "We were born here and we're staying here."
>
> Lai, however, says that the government is quite serious, and that the peasants soon will come to realize this. "They don't believe as yet because the money for the move is just starting to become available, and we have decided to prepare things for them gradually. They will accept it all later."
>
> "My people have lived here for a long time, as long as anyone remembers," Yang, 32, said earlier this month. "We don't see why we should move for some animals."

Wolong's peasants won the first round in this clash of wills. They refused to move into the new homes that were built for them with World Food Programme assistance, and most still stand empty. Instead, they built new homes with lumber from Wolong's forests to increase their government compensation in the event that they actually are resettled. Many of the trees that were planted as part of the program soon died of neglect and the bamboo was eaten by goats. And Wolong's population continued to grow, to over four thousand two hundred by the end of the decade.

When I left the project in early 1985, the crisis was not over. But then, the crisis will never pass because the panda's existence will always be threatened by something. I knew the species would survive the recent bamboo die-off. Little, however, had been done to control poaching, the most serious immediate threat to pandas, and habitat protection had only just begun. Nevertheless, the measures needed to protect the panda were clear; they only needed to be implemented. Ken Johnson planned to assist Chinese teams in censusing pandas, and Don Reid would soon join the research effort at Wolong and Tangjiahe. WWF expected to continue its participation in various ways, including the preparation of management plans.

So I left, filled with genuine hope for the panda's future. This hope was not a romantic illusion or the need to cover despair with optimism. Although serious problems remained, and the gap between policy and implementation was still great, the Chinese were showing concern and dedication in trying to help the pandas despite immense social pressures that make their task difficult. The resilience, tenacity, and pragmatism that have made the Chinese endure as a great people would, I hoped, be applied to the panda.

There is sadness in completing a project, in ending a phase of life to which one has been wholly dedicated. In my mind I would often return to all I had seen and enjoyed, carrying within myself the panda's world. I would often think of my many colleagues in Sichuan who had shown me the patience, kindness, and quiet support that with time and in another age could have ripened into long-lasting friendship: Reality would soon change into dreams. But at least I could still follow the panda's future through the eyes of others.

13

Prisoners of Fate

1985 to 1991

After leaving the project in early 1985, I maintained only tenuous contact with the pandas. I changed from participant to observer, reporter, commentator, critic, and protester, whatever the situation seemed to demand. I had assumed somewhat naively that a good panda conservation program was in the initial stages of being implemented. But I had forgotten that pandas attract adversity as readily as adoration.

Within a year I discovered myself once again on the usual roller coaster between hope and despair. I criticized with fervor whenever negligence threatened the animals. I continued to feel that it was my responsibility to help the pandas; I could not abandon them. Was the panda merely a prisoner of fate with a future that could not be altered or escaped? I read that Chinese researchers had announced that pandas are doomed because of a sperm defect, their small sperm making "impregnation very difficult, with little chance of successful fertilization." Fortunately pandas have no access to newspapers to create doubts in their mind, and they continue to reproduce well in the wild, as they have for millennia. But it would be tragic if as a result of false perceptions the pandas were relegated onto an evolutionary scrap heap.

Wolong and Tangjiahe had once been my home. I last saw the mountains in 1985, but when I left it was not really an end. There are too many memories and some regrets. Don Reid remained with the project in the field until late 1987, and Ken Johnson intermittently to late 1988. Through them I continued to participate vicariously in the life of the research and conservation efforts.

Most pandas I had known around Wuyipeng were dead; only Long and the ear-tagged male No. 81 were still alive in mid-1985. As these animals were not radio collared, they revealed little of themselves. The female Li was far outside the study area past Niutoushan, and only Hua's radio signals were being received. The project was in the doldrums. However, on 27 December 1985, a new female, Xin Xing, was captured and collared. And a year later, on 4 January 1987, Xin Yue wandered into a trap. Unlike all the other females that had been captured, she had a young cub, and everyone looked forward to observing this pair. But it was not to be.

Don Reid told me that Xin Yue lost her baby in March. Chinese researchers found the body, still warm, with most of its insides and a hindquarter eaten. Two pandas were fighting nearby. Circumstantial evidence clearly pointed to a panda as the killer, especially to radio collared Hua, who was in the vicinity. However, tracks of a golden cat were on a trail about three hundred feet away, and this animal was blamed for the baby's death by the Chinese.

Male lions, black bears, and other large carnivores have been known to kill young of their kind, and, distressing as it may be to implicate Hua in the deed, such behavior is apparently part of panda society too. Life, however, soon renewed itself. Xin Yue mated again on 22 May, unusually late in the spring, and she was attended by two males. She would not have come into heat so soon if her baby had survived.

At Tangjiahe, Don monitored both the pandas and bears, working with a new research team because Teng Qitao and others had been transferred. Tang, the radio collared male panda, continued to spend most of the year in his old haunts around the mouth of the Shiqiao valley, but in the summers he moved to the high ridges. There he died in July 1986, apparently of an abscessed jaw, an old wound of unknown origin that had caused bone decay. Xue, the female we had caught only a few weeks before Kay and I departed, retained her range in the Hongshi valley where, like Tang, she moved onto the ridges in summer and down low on the slopes in winter. In May 1987 she was found dead, the cause of her demise unclear, yet another of our acquaintances gone after a life much shorter than I had expected.

Don radio collared one more panda, one with but a brief history in the annals of the project. Xi-Xi was an adult female who was found in weak condition east of the reserve, apparently after being attacked by wild dogs. She was treated by a doctor in Qingchuan, radio collared, and released in the reserve in April 1985. Within a few days her radio signal vanished. From one of the highest peaks in the reserve, Don finally picked up the signal far to the west in Pingwu county, and always inactive. After a lengthy search, he found her dead at treeline. Xi had traveled in almost the opposite direction to the one that would have taken her back to her old haunts. That Xi, unlike Zhen in Wolong, neither headed home on being released nor settled down in a nearby bamboo patch is interesting, and worrisome. The task of resettling pandas into new areas may be more difficult than anticipated if they follow Xi's example and wander off.

Chong, the small male black bear, removed his radio collar when he awoke from his winter sleep in early 1985. However, Kui, the other male black bear, with whom we had lost contact in late 1984, reappeared in June 1985. During the last week of November he apparently entered his winter den, a hollow fir far up the Shiqiao valley near the Gansu border. His radio ceased to function in 1986. In July 1985, Don caught Dan-Dan, a yearling male black bear weighing a mere sixty pounds. Inexperienced in finding a good den site for the winter, Dan hibernated in a nest of bamboo and moss on a cliff ledge beneath an overhang, a drafty place exposed to the elements. He was radio tracked until October 1987, when his radio collar fell off, as it was designed to do when the bear's neck expanded with age.

Thus, by the end of 1987, the animals I had known, except for Hua, were either dead or their radios had ceased to function, shrouding their lives in mystery. After seven years, my involvement with individuals was over. And, as it turned out, research almost ceased too. Wuyipeng was virtually shut down by the spring of 1988. Hu Jinchu did little field work after I left the project, and other Chinese researchers made only short-term commitments. Radio tracking of pandas ceased. A WWF memorandum in 1989 noted, "At present there is essentially no research going on at Wolong." Our old research base at Tangjiahe was "deserted, quite desolate and depressing," as Jenny Johnson wrote us after a visit in 1988. Jiang Mingdao, Tangjiahe's dynamic leader, was soon thereafter transferred to another post, that of director of a gold mine, and his place was taken by someone passive. Patrols ceased and poachers promptly flourished. My colleague Teng Qitao also left the forest service. Two "rescued" panda young were shut away at the Tangjiahe holding facility under deplorable conditions, according to foreign visitors. We seemed to have left no legacy of scientific commitment to the pandas at Wolong and Tangjiahe.

At least Pan Wenshi, who had been an integral part of the Wolong program until 1984, has carried on our tradition elsewhere. Assisted by Lu Zhi and others, he has for several years now conducted panda research in the Qinling Mountains of Shaanxi Province. Several pandas have been radio collared and these have become used to the presence of observers. Consequently Pan Wenshi and Lu Zhi have been able to study the obscure lives of these animals as no one before them. They found, for example, that a mother may leave her cub alone in a den for as long as two days. Youngsters begin to scent mark at fourteen months of age. They may remain with their mother for at least twenty-two months, rather than just eighteen months as I had thought, even while she mates again. In 1988, Pan Wenshi and his colleagues published a book entitled *The Giant Panda's Natural Refuge in the Qinling Mountains,* an important contribution to panda biology.

Pan Wenshi and his team captured an unusual female panda in 1985. Her coat was a light chocolate brown where normally she would be black. I saw this animal, Dan-Dan, in the Xian Zoo and thought the aberration most unattractive. Mated with a normal-colored male, she gave birth on 31 August 1989. When it was about four months old, the infant's black fur changed to brown like its mother's, the *Xinhua* News Agency reported. There is now a danger that the freak gene will be perpetuated in the captive population.

Qui Mingjiang, the best of our young coworkers, represents the next generation in the project's legacy. He departed in mid-1988 for the United States to do graduate work in conservation biology, a fitting climax to his years of dedication. His training represents an investment in China's future.

What had the project accomplished after seven years of intensive effort? With an air of melancholy and resignation, Don Reid expressed it concisely in a memorandum to WWF:

> The biggest achievements have been in research: elucidating the fundamental behaviour, food habits and habitat requirements of the giant panda; and developing an understanding of the dynamics of bamboo growth and regeneration, and its interactions with the forest. We have demonstrated quite conclusively that poaching and habitat destruction rank highest as threat to panda survival, and that bamboo flowering can be accommodated by panda populations given sufficient space and diversity of bamboo species. Also, we have trained some Chinese scientists and technicians. We have influenced some of their thinking and motivated some of them to argue on behalf of wildlife and better conservation within their reserves. Now we are well underway in the development of Management Plans for the reserves and a Master Plan for giant panda conservation.
>
> Overall, however, we have little tangible evidence of success.

On 10 August 1986, Li-Li, a long-time resident at the Wolong research center, gave birth. Some claimed the baby was the result of artificial insemination, and others claimed she had mated naturally. No matter. It was an event worthy of celebration, and Lan Tian, Blue Sky, as the baby was called, became an instant media star, as reported in *WWF News*.

Lan Tian gazed out blearily on a world gone mad. Nothing in her short life had prepared her for the pandemonium that had suddenly hit her home in the heart of China's beautiful Qionglai mountains.

A few yards away photographers and television cameramen jostled to take her picture, ready to bounce it off satellites and to wire it around the globe. For Lan Tian—the first baby panda ever born at the Wolong Research Centre, built jointly by the Chinese government and WWF—has become a unique symbol of hope.

Certainly one hope was that the center's first achievement would soon be followed by others. But there were none, not in panda breeding and not in research. Lan Tian died in 1989 at the age of almost three years, apparently of intestinal problems.

Seven years after the research center was completed, it still had not attracted competent researchers, and it remained largely unused. One WWF memorandum noted, "The Hetauping Centre is neglected and rundown, and few attempts are made to increase the likelihood of natural breeding. . . . The general condition of the laboratories and the clinic was poor: both were filthy." Or, as one Chinese official told me simply in 1987, "The center is not working."

The deplorable situation at the research center became the focus of world attention in mid-1990. John Phillipson, a retired Oxford University zoologist, was commissioned by WWF to review that organization's work over the past quarter century. The Phillipson report was blunt on the subject of the Wolong research center:

> WWF has not been effective or efficient in safeguarding its massive investment. . . . WWF subscribers would be dismayed to learn that the capital input has been virtually written-off.

Prince Philip, president of WWF, was questioned by a reporter from England's *Sunday Express* (29 July 1990) about the Phillipson report and WWF's multimillion dollar infusion of funds for the panda, and he gave a candid answer:

Q. Does the WWF deserve criticism for spending money to little effect, to protect the giant panda?

That was very disappointing, one of those projects that was a good idea at the time. We thought, let's go in and help the Chinese protect the panda.

It resulted in long and often acrimonious discussions with the Chinese about how to do this. In the end, WWF agreed to put a lot of money into China which, in retrospect, we should never have done.

That was a long time ago and we were stuck with it and the Chinese demanded every single item on that agreement had to be met, in spite of the fact that they knew perfectly well that a lot of the items, like the elaborate equipment for instance, was nonsense. It was mortally embarrassing for them when they couldn't get people to use it.

I was very worried when I was there, because 80 percent of the bamboo forest had gone, largely through human encroachment. The protected areas were opened up and people moved in, cutting down trees and cultivating the land.

Then came Tiananmen Square, which threw the whole thing into confusion.

In retrospect, the project deserved criticism, but I think we would have come in for much more serious criticism if we had refused to make any effort to try to prevent the panda from becoming extinct.

Unfortunately, in spite of WWF spending a fortune on it, the chances of the panda surviving at the present rate of progress are not good.

Stung by WWF's criticism, Song Huigang of the China Wildlife Conservation Association was quoted as saying, "They have no right to interfere in China's affairs. . . . They use our panda as their emblem and have raised US $20 million. But only US $2 million of the money is put for use in China's panda rescue efforts."

Before the onset of this furor, the Ministry of Forestry had actually made an effort to increase the panda's chances of breeding at the research center by concentrating many of its captive animals there. About fifteen pandas were housed at the center by the end of 1990. In the spring of 1991, the Chengdu Zoo sent a male, a proven breeder, to Wolong. Given the usual reflexive antipathy between Chinese organizations, this was a notable gesture. I was told that two males mated with four females there that spring. Susan Mainka, a veterinarian from the Calgary Zoo, arrived in mid-year to assist with the breeding program for at least two years, the first time a foreign scientist had made a major commitment to the research center. On 7 September 1991, one female gave birth to twins and both survived, one cared for by the mother and the other by veterinarians. Although one of these young died in February 1992, the births lifted the aura of doomed hopes and futility which like a smog had shrouded the research center for years.

The Chinese Association of Zoological Gardens had in 1990 divided the country's zoos into six districts to improve cooperation in panda breeding. In 1991 this worthwhile effort resulted in seven pregnancies and the births of nine young of which five survived. But twenty-three females had come into heat, showing that the gap between the potential and actual number of births is still great.

The panda rescue work, a legacy of the 1983 bamboo die-off, continued well into 1987, long after there was any justification for it. According to official figures, sixty-two pandas were found dead. Of one hundred and eight pandas taken into captivity, thirty-three soon died, and thirty-five pandas were relocated soon after capture to areas with abundant bamboo.

I was concerned about the forty captured pandas that still languished in captivity, some sold to Chinese zoos, some residing in installations controlled by provincial forest departments. There were about ninety captive pandas in China in 1987—far too many—and these were not part of a coordinated breeding program. Hunched in corners of their iron-barred cages, most would pass their years viewed by an enthusiastic public that sees only a clownish face, not the haunting image of a dying species. Few of these pandas would ever breed. However, if most of those that were rescued after the bamboo die-off were given their liberty they would perhaps replenish the forests.

In 1987, China's zoo association began construction of a panda-breeding facility near Chengdu, one much larger than the one in Wolong. How would this facility be stocked? I have a nightmare vision of ever more pandas being drained from the wild until the species exists only in captivity. There should be a moratorium on capturing pandas. Yet as recently as 1991 eleven pandas were "rescued".

In the spring of 1987, WWF launched a major public appeal for funds to assist with protection of the panda's habitat. There is, of course, no doubt that the most serious long-term threat to the panda has been and continues to be habitat destruction, a fact recognized for years. However, the single most urgent problem facing the panda is poaching, which has for decades kept populations low in areas with ample bamboo. Poaching is relatively easy and inexpensive to control or at least reduce, but the necessary patrolling is a hard and thankless task with little glory or public acclaim. Chinese newspapers regularly report when a poacher has been apprehended and jailed. But even such public pressure has little impact on most local officials, and my private complaints seemed to have no effect at all. This surprised me, because criticism represents loss of face to a country that claims the panda as a national treasure, and lack of effort in this matter might turn world opinion against helping China's conservation effort. I hesitated at first to express my concern forcefully in public, fearing that China's response to criticism might affect WWF's work in the country. Yet, out of my concern for the panda, I ultimately had to be prepared to lose China's amity and to accept the risk of China's displeasure. The truth had to be widely known, for conservation must address all issues, not just the palatable ones, and the allocation of funds must receive the correct priority. The alternative—to keep silent in the face of serious neglect—was infinitely worse.

Wolong, as always, was a microcosm of the problems facing the panda as a species. In 1986, Ken Johnson and a Chinese survey team censused pandas in Wolong, the first such effort since 1974 when about one hundred and forty-five pandas were tallied. The result was a shock: only seventy-two pandas were found in 1986. Given the difficulties of estimating numbers on the basis of counting fresh droppings and bed sites, the figure is not precise. But there was no question that the panda population had declined drastically, perhaps by half, during a major conservation effort. One scientific report published in *Nature,* Britain's leading science journal, blamed the decline on the bamboo die-off. "As a result of the flowering of arrow bamboo in Wolong in 1983–85, 40–50 percent of its pandas may have died and others may have emigrated." But those of us who had worked in the field there knew better: the pandas did not starve, they were killed by poachers. Every year since 1981, the local administration had received our reports about snaring and complaints about lack of law enforcement. Yet poaching continued, even in our small study area. During one two-week period in 1987, Don Reid met hunters with guns and dogs and found two snare lines set for musk deer.

Poaching of wildlife in Wolong and elsewhere actually increased during the mid-1980s. The government's economic reform encouraged private initiative under the slogan: "It's glorious to be rich through hard work." *Ge ti hu,* private entrepreneurs, entered the animal-product market, hustling for quick profits. A state-run organization with the succinct name of China National Native Produce and Animal Byproducts Import and Export Corporation sent its agents into the countryside, further stimulating the killing of wildlife. Musk, bear paws, skins, antlers, bones, and other animal parts, often from endangered species, were blatantly sold in ever-increasing quantities on the open market (see Appendix A). And tragically, the panda was now killed not just inadvertently in musk deer snares, but deliberately for itself; the panda had become a commodity. Dealers offered poachers three thousand dollars or more for each pelt. The pelts were then resold as status symbols in Hong Kong, Taiwan, and Japan for prices that exceeded ten thousand dollars each. One survey in Japan during the mid-1980s recorded one hundred fifty-seven stuffed pandas and panda pelts. During 1986 and 1987, newspapers often reported panda killings. For example, three men and a woman were arrested with a panda pelt intended for illegal sale in

Guangzhou, and three men in Sichuan were fined and sentenced to prison for periods of from six months to one year for injuring one panda and killing another. Two panda skins were found in a Chinese vessel, sewn inside a quilt. A gang which systematically hunted pandas was smashed and twenty-six of its members were paraded handcuffed through Mianyang City in Sichuan.

Records kept by the Sichuan Forest Bureau show that almost half of the poached pandas are shot, a third are killed in snares and traps, and a few are first captured and then killed. In addition, seventeen percent are poisoned or bombed. Andrew Laurie, a British biologist who was involved with the panda project in the latter part of the 1980s, told me that local people in the Liangshan area attach explosives to meat bait to kill black bears, and pandas are also blown up.

One basic problem was that China still lacked comprehensive legislation to protect wildlife. The penal code merely stipulated a penalty of two years or less in prison for those who disobey hunting regulations. Generally the law applied only to those caught in the act, and it was seldom enforced against culprits who belonged to minority groups. Furthermore, the actual sale of fur and other parts of endangered species was not considered an offense.

In July 1987, a poacher in Pingwu county was apprehended after killing two pandas; he noted that the existing penalty of a two-year jail term was no deterrent. This helped prod the government into action. Shortly thereafter, the Supreme Court proclaimed that offenders illegally killing one panda, or smuggling one panda skin, will be sentenced to prison for ten or more years, life imprisonment, or even death.

That the death penalty has been instituted for killing a panda emphasizes, more than anything else, the crisis facing the species.

Yet poaching persisted. On 7 April 1988 the *New York Times* printed this news item:

> China has arrested 203 people for illegally hunting the endangered giant panda and recovered 146 pelts. . . . The Chinese Forestry Minister reportedly said 150 people still were sought in connection with an investigation into poaching and fur trading in Sichuan Province in central China. . . . Twenty-six people have been sentenced to prison terms ranging from a year to life, the Forestry Minister was quoted as having said. . . . Panda furs fetch high prices when smuggled to Hong Kong or Japan.

My reaction to the story was one of horror and sadness, natural feelings when anyone desecrates something beautiful. One hundred and forty-six dead pandas. Visions of a bloody rubble of bodies, and of the ghosts of Han and Ning and others who had strangled in snares at Wolong, rose before me. The magnitude of this destruction was such that it could no longer be ignored. For years I had railed against such poaching. However, both the Chinese and WWF, each for its own timid reasons, had preferred to brush the issue aside. Now, at least, poaching had become an accepted conservation priority. "Poaching is so serious," stressed WWF's Chris Elliott in a November 1988 news release, "that it poses an even greater threat to the giant panda than the bamboo die-off."

And the carnage continued in spite of drastic legal measures, including execution of offenders, as reported by the *China Daily* in October 1989

> Two farmers who were convicted of illegally selling giant panda skins were sentenced to death by the Intermediate Court of Mianyang in Sichuan Province on Wednesday. One offender, Liang Yongzheng, sold seven panda skins for 300,000 yuan [$81,000]. The other, He Guanghai, sold five panda skins for 30,000 yuan [$8,100]. Two of Liang's accomplices were sentenced to life imprisonment.

But the financial reward for selling a panda pelt was now so high that even the death penalty did not deter poachers. "I couldn't earn that much in a lifetime. Even though I risked my life, it was worth it," a poacher was quoted as saying to police in China's Public Security Newspaper. "If you hadn't caught me, I would have been rich."

The seventh National People's Congress helped the panda legally by eliminating an oversight that had made it an offense to kill pandas but not to sell their skins. On the first of March 1989, a comprehensive new wildlife protection law went into effect that prohibits capturing, killing, and trading in species under national protection. However, the law has so far had only a modest impact on the illicit trade. In 1990, WWF sent a conservationist posing as a Taiwanese buyer into China. During her travels she was offered two live panda cubs for one hundred and twelve thousand dollars and a total of sixteen panda pelts, as well as the skins of tigers, golden monkeys, and other rarities.

Between 1985 and 1991, Chinese courts handled one hundred and twenty-three cases of panda poaching and pelt smuggling and convicted two hundred and seventy-eight persons, three of them to death and sixteen to life in prison. But the killing continues, subsidized by the wealth of Hong Kong, Taiwan, and Japan.

With the pandas' situation deteriorating, WWF took several timely initiatives in 1986 and 1987. A three-month conservation training course for reserve leaders, administrators, and researchers was given by John Marsh of Trent University in Ontario and Gary Machlis of the University of Idaho. John McKinnon and Andrew Laurie, both from England, joined Chinese colleagues in preparing preliminary management plans for several reserves, including Wolong and Tangjiahe, as well as a master plan for long-term panda conservation. Andrew Laurie also taught the research staff how to monitor and census vegetation and wildlife. And in Wolong he took an anti-poaching patrol into the forest, hoping that by seeing a good example the local officials might finally respond with measures of their own. The three-day trip did not proceed as planned. Each patrol member arrived with his own porter loaded with a vast amount of food, including live chickens and bottles of warming *bei jiu*. Barely into the forest, the group required a long rest. When Andrew, frustrated and impatient with all this inertia, climbed ahead, the others panicked and rushed back home to send out a rescue team of villagers who found Andrew waiting on a ridge for his intrepid team to catch up. Poachers in Wolong remained safe. In 1991, WWF sent a United Nations volunteer, the Englishman Stuart Chapman, to Wolong for two years to help protect the forests against illegal woodcutters and panda poachers, a task the Chinese certainly could and should handle for themselves.

Robert de Wulf from Belgium gave a course in the interpretation of aerial photos and satellite imagery, important for mapping the distribution of panda habitat. Michael Baron and Nayna Thaveri from Hong Kong prepared portable education exhibits and brochures depicting the diversity of Wolong's plant and animal communities and panda conservation.

All this activity at least prodded Wolong into beginning its own environmental education program. A mobile education unit began to visit schools and villages, and slogans and notice boards conveyed conservation messages. Subsidies are now given to farmers under a contract system based on their compliance with the five nos—no fire, no cutting trees, no hunting, no ploughing, and no damaging forest regeneration—though laxness in enforcing compliance has been criticized.

With extraordinary tolerance for the burdens of disillusion, WWF still continues its commitment to the panda.

During the mid-1970s, many people took part in a panda census that covered all parts of the animals' range. The panda population was estimated at one thousand and fifty to one thousand one hundred. However, since census techniques were not standardized and teams were not trained, the accuracy of estimates varied greatly from area to area. The results for Wolong were, I believe, of the correct order of magnitude, but this was not true of Tangjiahe, where the original estimate of four hundred was later reduced arbitrarily by half, when in fact the actual number was probably below one hundred. An accurate, up-to-date population estimate was needed, one upon which a detailed, workable conservation plan could be based. The exact distribution of pandas and of the various bamboo species had to be plotted ridge by ridge and range by range. It was a major task, requiring at least two years of effort. I was pleased that Ken Johnson, with his wife, Jenny, was willing to join thirty-five Chinese to conduct the survey. And later Seiki Takatsuki from Japan briefly joined the team as bamboo ecologist.

The China–World Wildlife Fund Giant Panda Survey began work in September 1985. Taking one county at a time, the team hiked along compass bearings or transects across the mountains—during the first year alone they walked more than eighteen hundred miles—and counted fresh spoor, especially the number of rest sites. Pandas rest an average of two hours or more once or twice a day. With this information, transect results can be converted to approximate panda density. Droppings reveal food preferences, and stem length fragments in the droppings can indicate the age of a panda, whether it is an infant, juvenile, or adult. The composition, size, and density of trees and bamboos reveal the panda's habitat choice. Socioeconomic data on people in communes bordering the panda's range were also collected; however, WWF investigators were not allowed to see this information. Thirty of Sichuan's counties were surveyed, as well as nine in Shaanxi and one in Gansu. The censuses at first went well, and Ken was impressed with his coworkers' "ability to sacrifice and endure hardship for the common good" and the "cooperative team spirit." But all too soon, spirits flagged and work was often done with what can generously be described as a minimum of interest. Work as usual was not allowed to interfere with personal desires.

Completed after three years, the survey nevertheless yielded valuable information. Panda habitat is now so fragmented that animals are divided into twenty-four separate and isolated populations, some with fewer than twenty animals surrounded by cultivation and rugged terrain. Most counties have little bamboo forest left, often less than ten percent of the total area and none more than thirty-six percent. The remaining forests are being cut at a frightening rate. In Sichuan, suitable panda habitat, not all of it actually occupied by pandas, has shrunk to scarcely a half of what it was in 1974, and the situation in Gansu and Shaanxi is equally grave. Pandas are now found in only four thousand six hundred square miles of terrain.

The transects provided the basis for a preliminary estimate of panda numbers. The total: eight hundred and seventy-two to thirteen hundred and fifty-two animals. About half of the pandas were found in the Min Mountains, eighteen percent in the Qionglai Mountains, and most of the rest in the Liang and Qinling Mountains. Though seemingly precise, this estimate of panda numbers conveys only a general order of magnitude. There are many biases in collecting transect data, especially with unsupervised survey teams. As one report to WWF laconically noted, "there are numerous discrepancies between the instructions in the Work Plan and the records actually kept." Probably a minimum of thirteen hundred and fifty pandas still existed in 1988, but numbers have no doubt declined since then, due mainly to poaching. The annual rate of population decrease has been calculated at five to seven percent on the basis of rather fragmentary data. By coincidence, the population estimates for the mid-1970s and late 1980s were similar, each totaling about one thousand animals. But it is now known that the earlier estimate was much too low. The current results provided for the first time a realistic assessment of panda numbers, as well as of distribution, and upon this information an effective conservation plan could be based.

In August 1989, after a decade of cooperation with the Ministry of Forestry, WWF at last published a conservation strategy for the panda. The document is entitled "National Conservation Management Plan for the Giant Panda and Its Habitat" and was written by John MacKinnon, Bi Fengzhou, Qui Mingjiang, and five other Chinese. The plan provides an important blueprint for the panda's survival. Its stated objectives are:

1. To maintain in perpetuity a viable population of wild-living giant pandas as an integral part of China's Natural Heritage together with those plants and animals that together constitute the natural habitat of the panda.

2. To restore those areas of panda habitat that have been damaged or destroyed by human activities but are considered essential for the attainment of objective 1 above e.g. to maintain outbreeding between isolated populations and allow movement in cases of bamboo flowering.

3. To modify forestry operational practice in those areas where pandas occur outside the protected area system so that production operations in those forests are done in a way that is compatible with the survival of pandas.

4. To improve the breeding of pandas in captivity so that no more pandas need to be removed from the wild and extra captive-born animals can be used to strengthen the wild population.

5. To conduct the necessary research needed to better understand the biology, ecology and behaviour of the giant panda and the threats to the pandas, also to apply the most effective management to ensure the long-term survival of viable and ecologically stable populations of both the wild and captive pandas.

Based mainly on information collected by Ken Johnson and his Chinese colleagues on the survey teams, the plan discusses panda numbers, distribution, and condition of habitat in each of the thirty-four counties inhabited by pandas. It also notes that the fourteen existing reserves are too small to contain viable panda populations and proposes the establishment of fourteen new reserves totalling about fifteen hundred square miles. Eleven of these proposed reserves are in Sichuan, one is in Gansu, and two are in Shaanxi. The Wujia Reserve, one hundred and eighty square miles in size, is among the most important of these, because it would connect the Wanglang, Juizhaigou, Baihe, Baishuijiang, and Tangjiahe into one large protected area. Also critical is the proposed Minshan Reserve, nearly five hundred square miles of good panda habitat in an area of the Min Mountains that has as yet received little protection. Most other reserves are small, fewer than one hundred square miles each, but all provide various isolated panda populations with some hope of a future. In addition, it is proposed that the borders of several existing reserves be realigned. The Tangjiahe Reserve would be extended eastward, and certain portions of a number of other reserves would be excised to exclude villages and farmland. Where this is not possible, as at Jiuzhaigou, villagers would be removed and resettled.

The plan also discusses various realistic management options. These are worth quoting (see Appendix D) to reiterate what must be done and to emphasize that rather modest measures can provide the panda with security.

Not that the plan is without flaws. Political expediency rather than concern for the panda appeared to have dictated several recommendations and caused certain issues to be ignored. There is no discussion of reintroducing pandas to suitable bamboo habitat in provinces such as Hubei where animals do not now occur but did so in the past. Poaching, the most serious of the panda's problems, is mentioned, but the plan avoids the issue by failing to recommend detailed measures to control it. The sociological aspects of the panda crisis are simply ignored. Since local people poach and destroy habitat, they must be part of the solution to these problems. Conservation cannot be imposed from above. Any conservation effort must involve local people, based on their interests, skills, self-reliance, and traditions, and it must initiate programs that offer them spiritual and economic benefits. In-

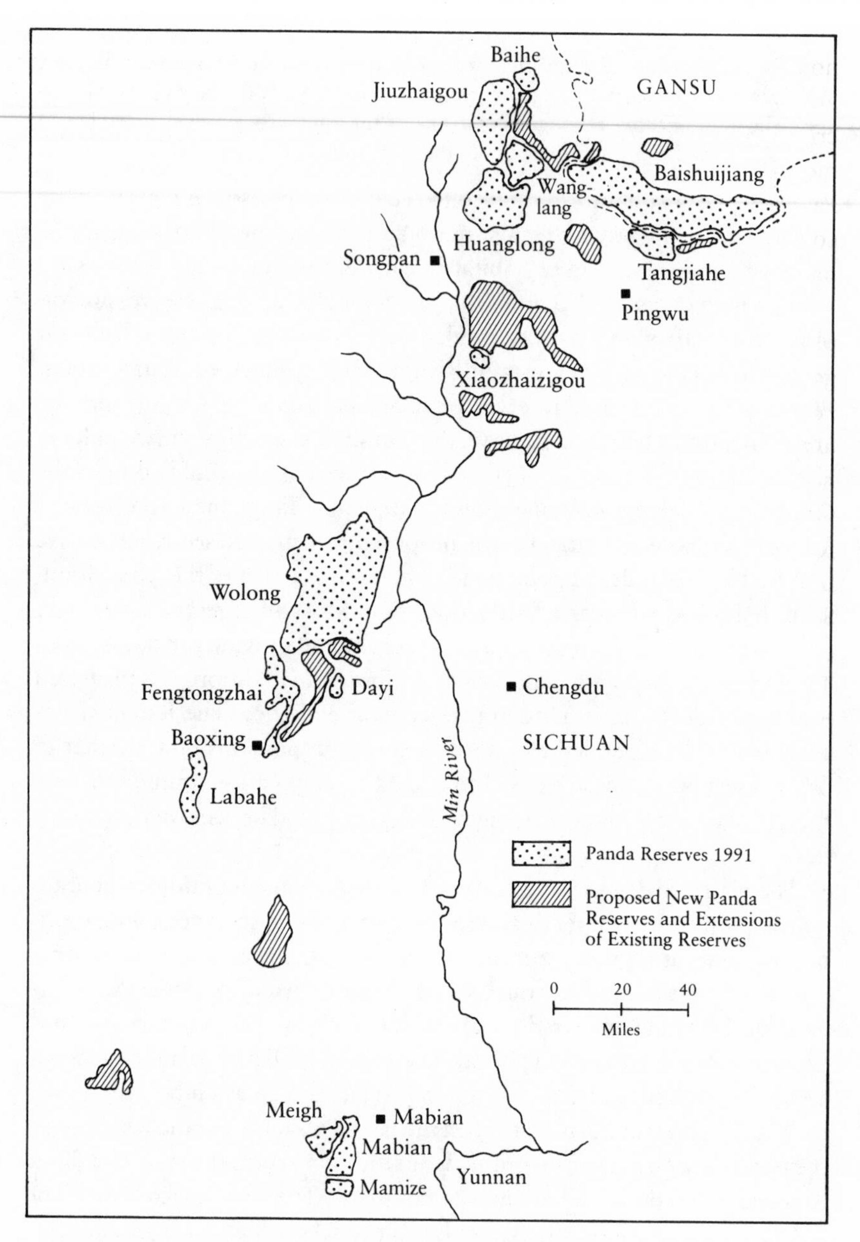

Existing and proposed panda reserves

novative programs of this kind have been developed worldwide in and around various reserves, some based on tourism, some on sustained use of critical resources. Among these reserves are Kenya's Amboseli, Rwanda's mountain gorillas in the Virunga Volcanoes, Costa Rica's Guanacaste, Papua New Guinea's Ubaigubi, and Nepal's Chitwan, to name a few that could provide helpful lessons and insights on how to integrate the needs of pandas and people.

There is, in addition, one other problem with the whole management plan. WWF published the plan before it had been ratified by the Chinese government. The plan's proposed budget was so high that the Ministry of Forestry anticipated problems with approval from the State Council. And after the government imposed austerity measures following the bloody upheavals of June 1989, the plan languished. It still had not emerged from the Chinese bureaucracy two years later. By then the Ministry of Forestry had revised it extensively. The budget had tripled to about fifty-five million dollars (at the 1991 official exchange rate) for the 1991–95 five-year plan. Several proposed new reserves were now considered useless after further evaluation, with too many people, too few pandas, and degraded forests. To keep the total number of new reserves at fourteen, as in the original plan, other proposed reserves were simply divided. Most of the corridors that were to be planted with bamboo and trees to link isolated panda populations were eliminated from the plan as being unachievable.

I gleaned this information from conversations in China; I cannot vouch for its accuracy. As of the end of 1991 only a Chinese version of the plan existed, and the Ministry of Forestry has not released it for translation, not even to WWF. However, at a meeting in Beijing during October 1990, WWF approved this new version. The plan was finally ratified, but not fully budgeted, by the State Council in January 1992.

A conservation strategy for the panda now exists, a crucial first step. The next step, one infinitely more difficult, requires the implementation of the plan. Over the decades and centuries, constant initiative, vigilance, cooperation, and dedication will be necessary to achieve the goal of prolonging the panda's tenuous hold on existence.

The panda project began with headlong passion for a grand cause, perhaps a romantic fantasy, and with an optimism that arose out of the vaguest understanding of the true dimensions of the problems ahead. The realization that the panda has so suffered and declined in numbers while we chronicled its life burdens me painfully. Enthusiasm and goodwill count for little when the enemy is a vast bureaucracy of local officials who myopically use obstruction, evasion, outdated concepts, activity without insight, and other tragic traits to avoid central-government guidelines and create ecological mismanagement on a dismaying scale. There is so far little protection of wildlife and forests, much less actual management of habitats for conservation. Segregated on its mountaintops, harried by poachers, its habitat shrinking, the panda has become an elegy. No face-saving illusions can hide these facts. Unless sound planning and vigorous law enforcement are soon initiated, all the field research, impressive laboratories, educational campaigns, public appeals, and legal assaults will be of no avail in saving the panda. To let the species slip quietly into oblivion would be hope's final betrayal.

Rent-a-Panda

No doubt, we need more foreign exchange, but this should not be at the cost of the life and liberty of the beautiful species in our country

Indira Gandhi
Prime Minister of India

Between 1936 and 1946, a total of fourteen pandas reached Western zoos, the result of a lively scramble to exhibit these rare and popular animals. The animal dealer Floyd Tangier Smith made a concerted effort during that period to obtain as many pandas as possible. His method was described in the notes of John Tee-Van, a former director of the New York Zoölogical Society, during a trip to China in 1941 to bring home two captured pandas:

> He began to advertise to the natives, putting up signs and offering rewards to hunters. He also established information centers along the route he travelled. He subsidized head-hunters who in turn paid farmers, medicinal herb gatherers, charcoal burners and any others whose business carried them far afield.

One wonders how many pandas died before reaching Mr. Smith. But the Second World War, followed by liberation in 1949, ended such exploitation.

Pandas attained a new status, that of official goodwill ambassadors, when in 1957 the Chinese government gave one to Russia. By 1983, a total of twenty-four pandas had been presented to nine countries as an expression of friendship, including two animals—Ling-Ling and Hsing-Hsing—given to the United States in 1972. This period of panda gifts came to an end in 1985, shortly after the bamboo die-off.

In 1984, China presented two pandas on short-term loan to the Los Angeles Zoo in conjunction with the Olympic Games. After three months, these pandas went to San Francisco, also for three months—but for a hefty rental fee. This commercial gesture precipitated much vigorous jostling by North American and European zoos to obtain pandas for exhibition, as well as by China to loan them. It became a rent-a-panda program, zoos vying for status, publicity, and profit by displaying pandas and China collecting six-figure fees for each loan, as well as a percentage of souvenir sales in many instances. Each loan also provided numerous Chinese with overseas trips. An advance team checked facilities. Delivery of the animals included delivery of one or more Chinese leaders to attend ceremonies that extolled pandas as goodwill ambassadors. A staff of Chinese remained at the zoo to care for the pandas throughout their sojourn. And finally leaders came to closing ceremonies to laud the bridge of friendship that had been built. With such incentives, it is understandable that Chinese zoos began to importune visiting foreign zoo directors to borrow their pandas. Pandas were now treasured more for their display value than for themselves; they had become big business.

Should pandas be rented? No wild animal is more popular with the zoo-going public. Delighted crowds even watch pandas sleep, as they do much of the time. Zoos are cultural and educational institutions whose exhibits make people aware of rare animals and the need to protect them, and pandas are, of course, superb at raising public consciousness. However, a good case can also be made against panda rentals. One can rightly argue that captive breeding in China should achieve sustained success before individuals are sent on stressful world trips during which their reproductive life is wholly disrupted. After all, China still removes pandas from the wild to augment a captive population that dies faster than it reproduces. Initially, I favored strictly regulated loans because I felt that an open-door policy with China was needed, a policy that would enable foreign zoos to cooperate with the Chinese and provide encouragement, knowledge, help, money, and, where needed, pressure to improve panda management in the wild and in captivity. But I changed my mind after observing the greed, politics, lack of cooperation, and undisciplined scramble for pandas that characterized the whole loan program.

From the beginning, China's central government attempted to coordinate all loans, adhering to CITES, the Convention on International Trade in Endangered Species, which tries to regulate the trade in rare animals and plants, but some municipalities bypassed regulations. Similarly, the United States Fish and Wildlife Service Office of Scientific Authority, which has to review and issue permits for the importation of the animals, found itself involved in panda politics and buckled under pressure. When there were irregularities in the import of two pandas to one American zoo and permits were not immediately forthcoming, the White House intervened.

Alan Cohen, an attorney, was quoted in an issue of the *New Yorker* (13 April 1987) saying that Mayor Edward Koch of New York "was relentless in his pursuit of giant pandas for the Bronx Zoo. You have to give him credit for perseverance. When he was in Beijing in 1980, he talked so much about getting a panda or two that whenever his name cropped up people there who didn't know anything else about him would say 'Oh, he's the guy who wants the pandas'." Unfortunately, the Bronx Zoo did not find the mayor's enthusiasm infectious; it felt that China should first develop an effective captive breeding program before sending the precious animals on tour. Nevertheless, in December 1986, City Hall notified the Bronx Zoo that two pandas would be available for exhibition the following spring.

Zoos in Los Angeles, San Francisco, New York, San Diego, Toronto, and Busch Gardens (Florida) all hired pandas between 1984 and 1987, and Calgary, Atlanta, Columbus, Portland, and Seattle were bargaining for pairs for 1988 and beyond. The Toledo Zoo and Detroit's Michigan State Fair, only sixty miles apart, competed vigorously, each wanting to be the first to host pandas in the Midwest. Such public figures as Nancy Reagan, Jimmy Carter, and George Bush became involved in the transactions. Japan had pandas on loan, as did zoos in the Netherlands, Belgium, and Ireland. Australia exhibited a pair of pandas for its bicentennial in 1988, and these animals then visited New Zealand. Even the Mexico City Zoo, which had received a pair of pandas as a gift in 1975, and since then has successfully raised four young, the best breeding record outside China, took part in panda loans. In exchange for a gorilla, two orangutans, and some other animals, it provided a panda to the Memphis Zoo for short-term exhibit, much to the annoyance of China, which apparently felt its monopoly threatened. Switzerland was a happy exception to this commercial frenzy: it banned all loans.

Something had gone awry. The 1930s were back in a different guise. Ruth Harkness—who started the current panda cult—then wrote: "You felt that every zoological society of any note whatever was sighing for a Panda, that life would not be complete until they had one." Attitudes had changed little, though prices had: in 1938 the New York Zoological Society offered only twenty-five hundred dollars for a panda, delivered at its Bronx Zoo. Fifty years later, not only did zoos still want these cuddly expatriates, but so did commercial enterprises such as a supermarket chain and a state fair.

Pandas were in danger of becoming merely show animals. Zoos were in conflict with the conservation role they had accepted as a mandate and were now expected to reflect. To be fair, some zoos, as New York and Calgary, did not want pandas, but their municipal governments decided otherwise. China was hurting its image in the conservation community by seemingly using the panda for mercenary goals. However, as one official from the China Wildlife Conservation Association explained: "There's no question of price. But we welcome a donation."

I had anticipated the trend in rentals, and in April 1985 had written the American Association of Zoological Parks and Aquariums (AAZPA) to suggest that certain guidelines for accepting short-term loans be prepared. I hoped that recently captured individuals would be returned quickly to the wild, that China would initiate a coordinated breeding program instead of keeping animals scattered and alone in many zoos, and that pandas with reproductive potential (including young) should not be exposed to the stress of a tour and be deprived of the opportunity to mate. The AAZPA took up the issue and produced excellent guidelines, covering everything from travel to housing, and urged its members to accept only animals that are sterile or past breeding age.

Ignoring guidelines about which they had not been consulted, the Chinese sent whatever pandas were readily available. The New York Zoological Society, for example, was shipped a wild-caught adult female who, according to the Chinese, was incapable of breeding. Shortly after arrival she came into heat. The male sent with her was too young to breed and so much smaller than the female that he was kept separate from her. The Chinese feared that she might injure him. The loan of these pandas was an embarrassment for the New York Zoological Society, which more than any other such institution has contributed to the conservation of pandas and wildlife in general.

When Peter Karsten, director of the Calgary Zoo, was offered a fourteen-year-old male and an eleven-year-old female on loan for the 1988 Winter Olympics, he cabled the Chinese: "We are unable to accept giant pandas for exhibit loans which are of the age and physical condition to reproduce. We insisted on this point in our negotiations in Beijing and were fully assured that this was also the policy of the Chinese authorities and, therefore, required no written agreement. . . . We truly cannot proceed with the loan without losing credibility, membership and face which would devastate our position as a reputable and responsible zoological garden." Faced with an unambiguous stance, the Chinese quickly agreed to send instead animals that met the guidelines.

As with everything concerning pandas, the issue of panda loans generated much emotion, especially since commercialization gave the whole program an aura of moral laxness. There would have been less of an ethical compromise if the AAZPA guidelines for loans had been strictly adhered to, and if all funds received by China were applied directly to panda conservation. The Chinese Scientific Authority felt "that at least some of the funds received might be available for conservation, but note that there is no independent assessment of benefits," as a WWF memorandum summarized the situation. Indeed.

Actually two separate Chinese organizations rent pandas, and each uses its funds independently with little attempt at coordination. The Ministry of Forestry is one of these, its program directed by the China Wildlife Conservation Association. The other organization is the Chinese Association of Zoological Gardens, which is administered by the Ministry of Urban and Rural Construction and Environmental Protection. Much of its money has been devoted to the construction of panda breeding stations.

For instance, one hundred and twenty thousand dollars in panda rental funds from the Antwerp Zoo and six hundred thousand dollars from the New York Zoological Society were devoted to the construction of the panda breeding station near Chengdu. According to one official, quoted in the *China Daily* in April 1987, "We'll gather to the base most of the 24 fertile pandas now in China's zoos and some in the field." There has so far been little indication that this station and the one nearby in Wolong, each belonging to a different ministry, will cooperate for the panda's benefit. A third panda breeding facility, this one under provincial jurisdiction, has also been planned for Chengdu.

In 1988, widespread concern over the rent-a-panda program precipitated resolutions by several organizations (see Appendix E). WWF at first approved loans—and indeed benefited financially from the loan to the Toronto Zoo—but now it wanted to ban all loans: "WWF urges zoos outside China and the Chinese authorities to cease their involvement in exhibition loans of giant pandas." Shortly afterward, in February, the World Conservation Union (IUCN), met in Costa Rica and suggested "that any exhibitions of giant pandas should only be adjunct to and completely compatible with an international captive breeding program."

In mid-February 1988, a panda task force appointed by the AAZPA met and prepared an updated position statement, one that continued to endorse loans but with specific guidelines. One guideline stated that "AAZPA members should only accept animals for short-term loans which are adult specimens physiologically incapable of reproduction, preferably adult males and preferably animals that are captive-born." It remained unclear, however, how the Chinese would determine, or even if they would bother to determine, that an animal truly could not reproduce. A further guideline stated, "any revenues specially generated from exhibitions should only be devoted to support specific aspects of the conservation plan for the species. . . ." The task force hoped that this guideline would reduce the commercial aura of loans by designating funds for specific panda conservation measures in China. In May 1988, CITES also supported panda loans in principle if these benefited conservation.

The United States Fish and Wildlife Service (FWS) showed renewed concern over the panda rentals in a letter dated 4 March 1988: "The Service will not be able to take final action on any application until there has been an opportunity to discuss further with counterpart agencies in China the effects of such loans on wild or captive breeding populations, as well as how funds raised from the loans would be used to enhance the survival of the species."

But could the FWS now live up to its ideals under political pressure? I wondered how the Chinese and the many zoos would respond to such resolutions. Would concern for pandas prevail over human self-interest?

By early 1988 at least thirty American zoos and other institutions had applied for panda rentals from the obliging Chinese. Politicians pressured the FWS to approve permits on behalf of their constituencies, yet a consistent import policy still did not exist. Indeed previous permits had violated the U.S. Endangered Species Act and CITES, both of which stipulate that import of endangered animals and plants must either be for scientific purposes or enhance the propagation and survival of the species. At one point the FWS asserted that if pandas had been caught in the wild before being listed by CITES as endangered on 15 March 1984, then no import permit was needed. This position made the panda's legal status unique in the United States, different from that of any other species, and it created a situation which threatened the ability of the government to regulate all imports of endangered wildlife. Animal dealers could then demand similar exemptions for every species, using claims about capture dates that could no more be verified than those for pandas.

Conservationists also argued that panda rentals were primarily for commercial purposes, another violation of the Endangered Species Act and CITES. The Chinese approved a panda loan to the Michigan State Fair in Detroit, a commercial venture. And zoos stressed the dollar value of pandas. The *San Diego Union* newspaper headlined "Zoo nets $4.4 million from pandas' visit." The Toledo Zoo publicly projected an income of three and a third million dollars from exhibiting a pair of pandas, and the Toledo Chamber of Commerce expected a seventy-seven-million-dollar bonus for the city.

Amidst the controversy, Fred Dunkle, FWS director, announced in March 1988 that no more permits to import pandas would be given until the Chinese demonstrated that loans were not harming the species and that rental fees were actually used for conservation. This policy was short-lived. Within weeks, despite widespread opposition, the FWS issued an import permit to the Toledo Zoo, thanks to the intervention of Congressman Delbert Latta of Ohio. However, when zoo officials returned from China with their two giant pandas, they also returned to a legal battle. World Wildlife Fund–US and AAZPA joined in a lawsuit against the FWS, charging that commercial trafficking in pandas violated the Endangered Species Act and CITES. They asked that the Toledo pandas be promptly returned to China. Not mentioned in the suit was information that the two pandas had been taken from the Wolong breeding station, a fact the Chinese had not mentioned and a potent point in support of the argument that loan animals were selected with little regard for the well-being of the species.

The Toledo Zoo swiftly countersued, asserting that WWF–US had interfered with its contract with China, and it asked for ten million dollars for loss of revenue and defamation of character. Toledo then publicly assailed WWF's suit as a publicity stunt designed to raise money. Ed Bergsmark, president of the Toledo Zoological Society, was quoted as saying: "WWF has used the panda as its logo; it's got 57 different patents on things, it collects revenues going from everything from panda gear to pajamas." And in an unsubtle pressure tactic, Wang Menghu, on a visit to Toledo, speculated that cooperation in panda research between China and overseas organizations might be affected by the suit.

On 17 June, the United States District Judge Norma Holloway implicitly acknowledged the commercial intent of the Toledo Zoo by granting a preliminary injunction that barred the zoo from collecting a special entrance fee to the panda exhibit. There had been a surcharge of two dollars per adult and one dollar per child. The pandas, however, remained in Toledo for five months.

Three days after the court ruling, the FWS denied an import permit to the Michigan State Fair, in spite of intense state lobbying pressure. The State Fair lost its income from crowd-pleasing pandas and China's Ministry of Forestry lost a guarantee of three hundred thousand dollars that had been designated for the moribund research center in Wolong, although this was hardly an ideal choice for scarce conservation dollars.

And on 24 June, the FWS announced the suspension of all import licenses for pandas. The moratorium would remain in effect until a new policy could be drafted, one that would have stricter controls on the many rentals that were "posing new, cumulative threats to the wild and captive populations."

The Chinese considered the furor a loss of face largely because they did not fully understand the legal, biological, and ethical reasons for the suspension, nor did they comprehend the internal politics of the situation in the United States. Yet, at the same time, several Ministry of Forestry officials with whom I spoke were pleased that loans had been temporarily halted because overseas pressure for more and more pandas caused dissension and strife within their system. To open a dialogue with China on the issue, Edward Schmitt of Chicago's Brookfield Zoo and then chairman of the AAZPA's Giant Panda Task Force traveled to Beijing in July 1988 and there met with the Chinese Association of Zoological Gardens. Fruitful discussions were held on the need to increase collaboration between Chinese and foreign zoos, the urgent task of preparing a panda studbook, and the desirability of drafting joint guidelines for panda loans.

When I had talked about Ed Schmitt's proposed visit with the the Ministry of Forestry and its affiliate, the China Wildlife Conservation Association, there had been interest in discussing the problem of panda loans with him, especially because the Michigan and Toledo fiascoes had involved the Ministry of Forestry and not its competitor in the panda rental business, the Chinese Association of Zoological Gardens. But when Ed Schmitt arrived in Beijing, the relevant Ministry of Forestry officials were unavailable. They had decided on their own political agenda, and soon arranged a meeting about panda rentals with the United States Fish and Wildlife Service. And the same ministry officials were still unavailable when they were invited by the Chinese Association of Zoological Gardens, the AAZPA, and WWF to participate in a symposium on strategies for breeding pandas in captivity. At the time the Ministry of Forestry controlled about twenty-seven pandas, and zoos in China about sixty. It is difficult to give a precise figure for the number of captive pandas because China had still not prepared a studbook to record all births and deaths, and provide up-to-date information on each living animal. The Ministry of Forestry probably avoided the symposium because a competing Chinese institution was also involved. Whatever the reason, the ministry might have learned something about improving the dismal reproductive record of Wolong's captive pandas.

Competition and lack of cooperation between the Ministry of Forestry and Chinese Association of Zoological Gardens severely hampered progress in developing a comprehensive conservation plan for the captive pandas. For that matter, the Chinese Association of Zoological Gardens could not even elicit much cooperation from its own zoo members at that time when it tried to improve conditions for breeding pandas. Of twenty-one Chinese zoos with pandas only ten had pairs. Is the panda a national treasure or the private property of those zoos and organizations that control animals?

The Ministry of Forestry and the Chinese Association of Zoological Gardens at least maintained enough contact to announce jointly in mid-September 1988 that all panda loans to the United States would be suspended. Just what the new policy might mean in practice remained unclear. For example, Zhang Yushan of the Ministry of Forestry was quoted as saying: "It's not that pandas absolutely can't go. It's just that such exhibits should be controlled." And a news item in Ohio's *Columbus Dispatch* stated: "The Chinese will keep their promise to Columbus and to Vice President George Bush that pandas will be here for the celebration [in 1992] . . ." The celebration is in honor of the five-hundredth anniversary of Christopher Columbus landing in the Americas. It remains unclear what pandas have to do with Christopher Columbus and his geographical misadventures. But, as *Newsweek* pointed out, it is perhaps supposed to illustrate "what Columbus might have seen if he had actually reached China." In any case, pandas could not enter the United States until the Fish and Wildlife Service issued import permits again.

Meanwhile China also continued its special brand of panda politics by offering the Taipei Zoo in Taiwan a gift of two pandas in October 1988. Taiwan at first demurred, because acceptance would seem to imply that the island was a province of China, the view of the Chinese government, rather than an independent country, as Taiwan sees itself. China had publicly ceased giving gifts of pandas to foreign countries. However, with an election near, certain Taiwanese legislators promoted the panda gift, hoping that the animal's popularity might reflect on them. In April 1989, Taiwan was ready to accept the pandas but "would not receive the precious animals immediately for fear of inadequate conditions to rear them," as one newspaper delicately put it. The political panda minuet continued when the Olympic committee of China took charge of the gift and offered the pandas to the Olympic committee of Taiwan. Finally, in 1990, Taiwan decided not to accept the pandas, a decision acclaimed by conservationists.

In spite of various problems, an acceptable loan policy was evolving, slowly and with detours. By the end of 1988, the Chinese Association of Zoological Gardens had drawn up its regulations governing the exhibition of giant pandas, and other Chinese organizations were also working to resolve contradictions in their policies. For example, only captive-bred animals or animals caught in the wild before 1983 would be sent on loan. All loan approvals would now be the responsibility of the Ministry of Forestry.

The United States Justice Department agreed to pay WWF–US one-third, or fifty thousand dollars, of WWF's legal fees to settle the lawsuit against the FWS, and WWF accepted the offer. This suit had challenged the importation of the two pandas to the Toledo Zoo. And the Toledo Zoo agreed to drop its suit against WWF and the AAZPA.

But Gong-Gong, the panda star of the Great Circus of China, prevented the furor about pandas from subsiding. During autumn and winter of 1988, he toured Canada, participating in about one hundred shows. It was clearly a commercial activity, yet the Canadian Wildlife Service, which administers CITES regulations in Canada, had issued an import permit. In its defense, the Canadian Wildlife Service noted that one million dollars was being donated to panda conservation in China. Critics in turn responded that the money might not be used for its intended purpose, the construction of a conservation and education center in Wolong.

Meanwhile, the London Zoo, Cincinnati Zoo, and Mexico's Chapultepec Zoo provided an imaginative example of cooperation in captive panda management. Chia-Chia is Great Britain's only panda, a male that arrived there in 1974. The Chapultepec Zoo has two young females and a young male, all born there and the offspring of the same parents. With inbreeding between fathers and daughters or between siblings obviously undesirable, another male had to be found for a breeding loan. London obliged. While in transit from Great Britain to Mexico, Chia-Chia had a three-month stopover in Cincinnati to raise funds before heading south in late 1988. The money helped expand the panda facility at the Chapultepec Zoo, and also created a panda conservation fund that will be managed jointly by London and Cincinnati.

Chia-Chia became a father for the first time at the age of about eighteen years when his Mexican mate, Tohui, gave birth to a cub in July 1990. But the tale does not end here. During a visit to China in 1988, former British prime minister Edward Heath asked China's leader Deng Xiaoping for a female panda to provide a mate for Chia-Chia, whose companion Ching Ching had died in 1985. China finally agreed to the request in mid-1990, and promised to send a female to the London Zoo in early 1991. But by then Chia-Chia had contentedly emigrated to Mexico on extended loan. Instead, the London Zoo planned to obtain another male from Mexico, seven-year-old Liang-Liang. All was set for another round of high-level panda diplomacy. However, in April 1991, the London Zoo announced that it would close by autumn unless it received a multimillion dollar infusion of government money to offset operating losses. This should have halted the panda transaction. But no. In October 1991, even though the London Zoo's financial crisis remained unresolved, China shipped an adult female named Ming-Ming to England on a two-year breeding loan. And Chia-Chia, still in Mexico, selected that moment to die. However, the story may yet have a happy ending. The Berlin zoo sent its male panda, whose mate had died several years before, to London for breeding.

The United States ban on panda imports in 1988 did not of course include other countries. Two pandas were hired for several months during 1989 by Winnipeg's Assiniboine Park zoo. The rental money was once again used by China to support a panda breeding station, the one in Chengdu. If the millions of dollars that have been raised from loans were spent on anti-poaching and forest protection measures instead of on the construction and maintenance of walls around pandas, the future of the species would be brighter. But, on a positive note, Australia's panda rental fees were used to create a new reserve in Gansu. Two pandas also went to Kofu city in Japan. Visitors were fewer than expected, and Kofu required all public employees to buy tickets to the exhibit. Pandas in Japan are sometimes sublet to other institutions in the country, assuring huge profits. Emulating the U.S. frenzy of 1988, Kumamoto and other Japanese cities also scrambled for pandas.

In 1990, WWF, IUCN, AAZPA, and the International Union of Directors of Zoological Gardens were finally out of patience with zoos and other institutions that in a greedy and unseemly manner disregarded the pandas' welfare. They voted for a worldwide temporary moratorium on all panda loans.

By the end of 1990, after two and a half years of deliberation, the United States Fish and Wildlife Service had still not revealed its new panda loan policy. Cynics predicted that the policy would surely appear in ample time for the Columbus Zoo to receive its pandas in 1992, as promised by George Bush when he was still vice president. The new guidelines were published in the *Federal Register* on 14 March 1991. They elicited immediate and vehement protest from the conservation organizations whose policies and resolutions had been ignored by the government. The new policy was so weak that virtually any institution could claim pandas if it had the money and if the animals were not "used for primarily commercial purposes." There was no demand for a world registry of captives, no demand for a global strategy to protect the species, no demand for accountability of conservation funds. Peter Karsten, then chairman of AAZPA's Panda Task Force well expressed his concerns in a letter dated 11 April, 1991:

> The policy by the USDI/FWS will cause a renewed flood of loan applications, which not only will put animals of this very precious gene pool into jeopardy to reach optimum breeding opportunities, but seriously undermines the conservation strategies that have just been formulated and had international consensus for strategies to preserve the species. This policy is truly sabotaging the very fragile infrastructure that was emerging to develop a comprehensive management plan for Giant Pandas in captivity and the wild. It is difficult for me to understand how an agency responsible for the conservation of natural resources can make such an uninformed decision, which I could understand if said agency was responsible for economic development and tourism.

In the absence of an acceptable government loan policy, the AAZPA now used its powers to try to regulate panda imports among its North American zoo members. Zoos which attempted to obtain pandas faced loss of accreditation. This threat effectively deterred the Burnet Park Zoo in Syracuse, New York. But the Columbus Zoo claimed an exemption because it had submitted its application for pandas before the AAZPA moratorium. Similarly, the International Union of Directors of Zoological Gardens passed a resolution for a moratorium on loans in August 1990. But Singapore exhibited pandas in late 1991 on the basis of a prior agreement with China.

The IUCN held its general assembly in Australia from late November to early December 1990. One of its resolutions urged that the panda moratorium remain in effect until July 1991. By then the recommendations from a major panda conference, to be held in Washington, D.C., would be available. The National Zoological Park planned to host this international conference, entitled "The Panda, a Conservation Initiative," in early June. The main purpose of the conference was to prepare a global breeding strategy and action plan for the species, including all animals in captivity. The conference failed in its primary objective. There was no agreement on a global conservation plan. There was not even a consensus on the criteria for selecting appropriate loan animals. Should subadults be exposed to the stresses of overseas trips? Should only males be rented and females left to reproduce in peace? Since pandas are solitary in the wild, why are they exhibited in pairs, a practice that doubles the number of animals on tours?

When the conference failed to produce a global panda strategy, the various organizations reaffirmed their stand on a moratorium.

On 14 August 1991, the AAZPA issued this statement: "As has been done during several of its previous meetings, the Board of Directors voted unanimously to continue in force its moratorium on the importation of giant pandas by AAZPA members."

And Martin Holdgate, director general of IUCN, wrote to CITES in a letter dated 17 July 1991:

> In the light of the failure of the meeting to conclude with this comprehensive strategy, IUCN has concluded that there is now no option but to recommend strongly that a moratorium on short-term exhibit-only loans be enforced worldwide. IUCN has concluded that these *non-breeding* loans are proving a distraction to the real business of establishing a conservation strategy that will save the species from extinction. We therefore urge that you implement a total *moratorium* on importation of giant pandas for *short-term exhibit-only loans* until a satisfactory conservation strategy for both captive and wild populations of the species is developed. However, we consider that breeding loans of giant pandas could be useful as a step to develop the necessary cooperative management programme for the captive population, and this should integrate well with the new studbook for the species that has been assembled by the Chinese Association of Zoological Gardens. . . .

The whole panda situation had become such a miasma that the moratorium had become essential. As I first wrote this, I wondered whether in 1992 the Columbus Zoo would forego profits for the sake of the panda's future. But I should not have wondered. Greed triumphed over morality, as it so often does. Ignoring the widespread opposition of conservation organizations and even the AAZPA, the Columbus Zoo continued to negotiate with China's Ministry of Forestry for a pair of pandas. In this the zoo was encouraged by President George Bush, who wrote Dana (Buck) Rinehart, mayor of Columbus, on 2 August 1991, that "Columbus is very special to my family because by grandfather lived and worked there for many years . . . ," and that the pandas would create "sustained good will and friendship between the United States and China." On 29 October 1991 (U.S. time; China time is a day later), a delegation from the city of Columbus and the Columbus Zoo signed a loan agreement with the Chinese in Beijing. Immediately afterward, on the first of November, President George Bush sent a handwritten note of congratulations to the mayor of Columbus: "Dear Buck, got your message. I'm so glad the panda deal is closed.

Great going—George Bush." With such an endorsement, the required permit from the FWS would present few problems. However, the proposed loan animals did not fit import guidelines in that both were adult and probably breeders, caught in the wild in 1986. And the application was incomplete in that it gave little indication of how China would spend the rental money to benefit panda conservation. The FWS was ready to deny the import permit. But then Congressman Wylie of Ohio contacted Secretary of the Interior Manuel Lujan who contacted the FWS, and the unsatisfactory permit application was suddenly considered proper after all. Subsequently, China agreed to send two young male pandas, born in captivity in 1989, instead of the adult pair, and the money derived from the loan was designated for the establishment of a new panda reserve in the Qinling Mountains of Shaanxi Province. The application from the Columbus Zoo now essentially met the FWS criteria for a permit.

While the world conservation community waited to see if the FWS would join in the moratorium or give a permit in response to political pressure, the Columbus Zoo had other problems, self-inflicted. Jack Hanna, the director of the Columbus Zoo, was brought before the ethics committee of the AAZPA where he was told that the zoo would lose its accreditation and face suspension for one year unless it cancelled the panda loan agreement. WWF-US planned to ask the courts for a restraining order to prevent the pandas from being sent as soon as the FWS issued an import permit to the Columbus Zoo. But in a preemptive strike, the Columbus Zoo went to court and demanded an injunction to prevent WWF-US from interfering. It also asked for one million dollars compensatory and punitive damages, as well as a trial by jury in Columbus, and all this before WWF-US had taken any action. At issue, of course, was the huge commercial benefit to the city of Columbus if it could exhibit pandas. It was calculated that Toledo derived a profit of about sixty million dollars from its panda loan in 1988, largely as a result of increased tourism to the city. (China's profit from a rental is a few hundred thousand dollars, a poor bargain indeed.) The court granted the injunction, but WWF appealed to the 6th Circuit Court of Appeals, which sent the case back to the lower court, noting that the zoo's allegations "appear very thin." On 20 April 1992, the FWS had issued the import permit to Columbus, and the pandas arrived in that city late in May. The AAZPA promptly suspended the zoo. WWF and Columbus then settled their legal dispute. The zoo agreed to donate $65,000 to WWF and to increase its contribution to panda conservation; WWF withdrew its opposition to the panda rental. And the FWS announced that it would once again evaluate its panda import policy. So the situation in mid-1992 was once more much as it had been in 1988.

I am confident that captive panda management will soon improve. China will publish a studbook; it will achieve an effective breeding program within the country; it will collaborate with institutions elsewhere in a worldwide panda propagation effort. Panda loans will resume under strict international supervision. Some day China will probably make as much or more money from panda tourism as from panda loans. The research in Wolong and Foping has shown that pandas readily become used to people. Visitors would certainly pay high fees to view pandas, just as they do to observe mountain gorillas in Rwanda and Zaire.

The politics and greed, coupled with the shameful indifference to the panda's welfare that has characterized much of the rental business, will not vanish. But I fervently hope that it will be contained. There are not enough pandas, nor will there ever be enough, to provide animals for all those who clamor for them, whether on loan, as a gift, or through purchase. If we want to burden the panda with symbolism, reverence, and adulation, fine. However, we also have a moral obligation to maintain the species in the wild. With panda numbers dwindling year by year, not every zoo, not every country, can have them. The panda has not evolved to amuse humankind.

Epilogue

I have endeavored to write this book as a forthright, honest account of the panda project, a biologist bearing witness, presenting facts, seeking truth; I hoped to convey the atmosphere of those years in China and sought to explore the life of a wondrous creature. It was certainly simpler to write about pandas than about the people and institutions involved with the project. On rereading these chapters, I note that at times I seem to air grievances or have become a public scold. That is not my intention. Some passages admittedly convey an impression of a mission gone sour, of the desperate disappointment I felt and still feel. However, a chronicle such as this cannot evade the duty of making judgments; the demons of memory, even if recalled in tranquillity, demand expression. Candor may not be appreciated, especially not in an authoritarian state such as China where penalties for dissonant opinions can be swift. I know what seems to be the truth, but according to Asian custom I should not speak it.

As I finish writing this book, over a decade has elapsed since the panda project began. Perhaps I should have published my account of the panda's plight earlier. But deeply involved with a wildlife project in Tibet, I hesitated to publish, afraid that China's government would terminate my research. Actually, conservation needs an organization patterned after Amnesty International, the human rights organization, to which an individual could quietly report acts of environmental vandalism and corruption without fear of retribution.

There was another reason, a more important one, for delaying publication. The project gradually developed an aura of disillusionment, of a vision crushed by reality, of a bitter wisdom that the panda was doomed. The panda seemed to have become a metaphor for hopelessness. Although I may express myself in terms of moral indignation and pessimism, I require hope and optimism. It is the difficult fate of this generation to finally grasp the magnitude of all the offenses against the panda and other forms of life;

the extent of environmental destruction has been nothing less than a spiritual divestment, a renunciation of past and future. I had wanted to write not just a nostalgic book about the decline of yet another animal but to describe a painful past balanced in the end by a proclamation of hope, a parable of sin and redemption. So I have waited, my optimism seemingly independent of logic.

That the panda has been enshrined as an icon of our environment is not surprising. The animal has the power to touch and transform all those who gaze upon it; it has only to appear to brighten a scene. Yin and Yang are the two great Chinese forces of separation and unity: black and white, dark and light, sun and moon, summer and winter, life and death. Each force carries part of the other, each needs the other to retain a balanced whole with an emphasis on suppleness and endurance. The panda personifies this Yin and Yang. But humankind has upset the balance, and the panda's existence is now shadowed by fear of extinction.

I can only view with irony the fact that never has the panda's destruction been as rapid as during the years we studied it, during a period when it received more attention than at any time in its long history. Our years of intensive effort to protect the panda have certainly not yielded a victory and perhaps only a modest postponement of defeat. Indeed I am haunted by the realization that the project may have harmed rather than helped the panda. Many persons and several institutions have genuinely had the panda's interest at heart and their good intentions are unquestioned. But had the panda remained in the obscurity of its bamboo thickets, free from worldwide publicity and the greed this publicity helped to fuel, there might not now be so many captives, needlessly caught during and after the bamboo die-off, and not so many breeding stations. Pandas might not have huge prices on their heads and be sent alive overseas as rentals or sold dead as grisly trophies to Taiwan and Japan. However, even if pandas had been left undisturbed, their peace would soon have been shattered by loggers and an ever-growing population that turns mountain slopes into fields. The project provided for the first time basic information about the panda's obscure life, it defined conservation problems, and it proposed solutions to these problems. That is the project's important legacy. The continuing decline of the panda must not cripple our sense of purpose, our will to save the species. The panda survives, as does its habitat, and a realistic plan to save the animal exists. But unless we act *now* to implement the plan, our brief pang of love for the panda will end with an eternity of remorse.

For pandas there exists no freedom other than the peaceful security of bamboo shaded by forest. It cannot adjust its existence to ours, it cannot compromise its needs. To think otherwise, to continue gambling on its future, would be a costly blunder. All our fine sentiments and humanitarian concerns, all our attempts to immortalize the animal while it still exists, are of no consequence if the panda is allowed to vanish. To provide assurance that the panda will remain as a perennial witness to the wonder of evolution and as a talisman of ecological redemption, we must make a global commitment.

The ultimate responsibility for saving the panda in its natural home rests with China: it alone can implement the measures needed for the animal's protection. The rest of the world must, however, offer guidance, funds, and moral support. The gravity of the situation represents both hope and opportunity. But if we fail to make the correct choices now, the last pandas will disappear, leaving us with the nostalgia of a failed epic, an indictment of civilization as destroyer. We cannot recover a lost world.

The panda has no history, only a past. It has come to us in a fragile moment from another time, its obscure life illuminated through the years we tracked it in the forests. This book presents that moment rather than a memorial.

My years of involvement with the panda permeate my soul just as a panda defines and fills the bamboo forest with its luminous presence. Although they shuttle between heartache and happiness, my memories dwell mainly on Wei and Tang and the other free-living pandas who touched my life. The image of Zhen lingers most in my mind. I remember Zhen vividly on a day when I left Wuyipeng, when her brief appearance seemed like a parting gift.

I had found her not far from camp sitting hunched on a moss-covered boulder, muzzle tucked into her folded arms. Quietly I approached to within forty-five feet and waited. She raised her head and with a disinterested gaze looked at me, then leaned forward, her back to me, and continued her rest. There was a startling self-assurance, a striking kind of freedom, in the way she ignored me. At intervals she changed position, resting on side or belly, and occasionally she sat up to scratch or paw flies off her face. Once she glanced in the direction of loud voices from camp. After two and a half hours, with the onset of a heavy rain, she raised her arms above her head, stretched, and yawned cavernously. She descended from the boulder and began to munch shoots. I left her there, her pelage gleaming softly in the bamboo twilight, until like falling snow she melted gently into the forest.

APPENDIX A

In Search of the Kylin

THE ENDANGERED WILDLIFE OF CHINA

Knowledge of the constant is known as discernment. Woe to him who willfully innovates while ignorant of the constant.

Laotse

Two Chinas exist, the eastern half where most people live, and the sparsely settled western deserts and highlands. In the east, I have traveled by train from the city of Chengdu northeast to Harbin, a sixteen-hundred-mile journey. The landscape is wholly tame, cultivated for millennia except for a few barren hillsides. All original stands of forest are gone. It is rare to see even a bird. In 1958 a national campaign against the Four Evils—flies, mosquitoes, rats, and sparrows—eliminated not just sparrows, which were blamed for eating grain, but other birds as well. Air rifles and slingshots have continued to keep bird populations low. And the export of birds for pets and food is enormous, three million in 1985, excluding the illegal market.

Even remote parts of western China have not been spared the destruction of wildlife. Rangelands in Xinjiang, which according to old accounts teemed with goitered gazelle, now lie lifeless in the heat haze. Little is known of the few hundred wild Bactrian camels that survive in the remotest parts of Xingjiang's Taklimakan desert. Mongolian gazelles once migrated in huge herds back and forth across the international border between Mongolia and China's province of Inner Mongolia. "The entire horizon appeared to be a moving line of yellow bodies and curving necks. . . . Thousands passed in front of us," wrote Roy Chapman Andrews in 1932. Only remnants of these vast herds survive, and China slaughters these. Between 26 October and 17 November 1989 about four thousand gazelles were shot in one area of Inner Mongolia, according to the *China Daily.* In many mountain valleys only an occasional argali sheep horn revives images of the past, and the crags above seldom echo to the clatter of ibex hooves or feel the soft pressure of a snow leopard's paw. With the advent of roads, high-powered rifles, and good markets for meat, skins, and horns, wildlife on the Tibetan Plateau has also been locally eradicated or decimated. Species such as wild yak, Tibetan antelope, and Tibetan wild ass are now largely confined to the bleak and sparsely inhabited parts of northwestern Tibet (see map, p. 150).

Most large animals in eastern China were displaced long ago and, if they survive at all, they cling to just a vestige of their former range. Sumatran rhinoceroses, Malayan tapirs, elephants, and gibbons, for instance, once lived as far north as the Yellow River in the tenth century. Now the rhinoceroses and tapirs are extinct in China and the elephants and gibbons survive only in scraps of rain forest that remain in Yunnan. In addition, about a dozen gibbons inhabit Hainan Island, the last of a distinct subspecies. Père David deer, once a common inhabitant of marshy river flats, survive only in captivity. The giant saltwater crocodile is gone from China's estuaries. Mongolian or Przewalski's wild horses and saiga antelopes disappeared from the grasslands of western China during this century.

China currently recognizes a total of ninety-eight mammal, bird, and reptile species as requiring full government protection. All are endangered because of habitat destruction, unrestricted hunting, or both—the two basic issues plaguing not just China but all countries—and several of these species are seriously threatened.

The *baiji,* or Yangtze River dolphin, may be one of tomorrow's extinctions. Scattered in small groups over miles of murky river, perhaps only two to three hundred *baiji* remain. Their numbers continue to dwindle as some animals are caught on large, bare hooks that are set in long lines by fishermen, whereas others collide with riverboats or succumb to pollutants, according to an article in *Whalewatcher* by Bernd Würsig. A planned high dam in the Yangtze's Three Gorges area would eliminate more than a hundred miles of the *baiji*'s remaining river habitat. A desperate attempt will soon be made to maintain a few animals in an old oxbow lake beside the river. Two institutions, the Institute of Hydrobiology in Wuhan and Nanjing Normal University, are leading the efforts to save the graceful *baiji* in the wild.

The tiger, however, has no such friends, because the people see it as a cruel and belligerent animal. Of the five tiger subspecies in China, one is extinct—it once occurred in marshes and along rivers in Xingjiang—and the others may soon disappear. The most extensive forest tracts in China are in the northeast. That area made ecological world news in 1987 when the Black Dragon fire swept over eighteen million acres, an area the size of Scotland, and also blackened twelve million acres on the Russian side of the border. (The great Yellowstone fire of 1988 affected a million and a half acres.) Siberian tigers were once abundant in these northern forests, preying on sika and red deer and on wild pigs. Excessive hunting has killed all but perhaps thirty scattered individuals. But in neighboring Russia the subspecies is at present moderately well protected and perhaps as many as three to four hundred animals persist.

Even sadder is the fate of the South China tiger, a wholly Chinese subspecies, which once roamed over fourteen provinces and was estimated to number four thousand as recently as 1949. In 1959, the tiger was declared a pest, together with the Asiatic black bear, wolf, and leopard. People were encouraged by the government to exterminate the cat "by all means." They had almost succeeded by 1977 when the South China tiger was given protection. Today, still hunted, perhaps thirty to fifty tigers persist in the wild, an estimate based on a survey made by Gary Koehler on behalf of WWF during the winter of 1990–91. The remnants are scattered in forest fragments over five provinces, Jiangxi, Hubei, Hunan, Fujian, and

Guangxi. The only serious effort on this tiger's behalf is a captive propagation program at the Chongqing Zoo. There are about fifty South China Tigers in zoos, more in captivity than in the wild. Once the last tiger in China has been sequestered behind bars, the species will never see the freedom of a forest again. Who will tolerate these large predators? If China wants free-living tigers, its first priority must be to protect those it still has.

The rarest species in China, and one of the rarest in the world, is the crested ibis, a striking bird with white plumage, a red face, and long, curving bill. Once the crested ibis was found all over eastern China, Japan, and Korea, but hunting reduced it to such an extent that until recently it was thought to be extinct in the wild. In 1981, however, seven birds were discovered in the Qinling Mountains not far from Xian. Their final refuge was several ancient oaks in a graveyard in which they nested and from which they foraged in nearby rice paddies. Strictly protected by the government and closely monitored by dedicated scientists, the crested ibis increased in number to seventeen in 1984, and there are at least forty today.

Chinese alligators, or "earth dragons" as they are called locally, endure only near the mouth of the Yangtze River in Jiangsu Province where fewer than five hundred remain, their range contracted more than ninety-five percent in recent times. Alligators can increase rapidly in number if they and their nests are protected, as American alligators have demonstrated. But in China, land reclamation has destroyed most alligator habitat. After surviving millions of years, the species now exists precariously in reservoirs and rice paddies. Chinese alligators were bred in captivity as early as 700 B.C. Efforts since 1983 to raise them in captivity have also been successful, with over six thousand hatchlings so far at one Chinese breeding farm. But this farm has now curtailed the breeding program because it does not know what to do with all its alligators. There is no adequate place to release them into the wild and CITES regulations prohibit the sale of these endangered animals outside the country.

Many other animals are endangered—the one hundred and twenty Eld's deer on Hainan Island, the few hundred white-headed langur monkeys in Guangxi, the remnant populations of Sichuan hill-partridge. These fragile numbers—forty South China tigers, forty crested ibis—are the last of their kind in the wild. Perhaps these numbers are enough to maintain the species, perhaps not. Whatever their future, these endangered animals symbolize the uncontrolled hunting and forest destruction that have eliminated most of China's wildlife.

Even obscure species in China suffer from the Asian predilection for treating wildlife as recipes and medicines. Slow loris, a rainforest primate, is best cooked with lemon leaves. To warm your blood in winter, cook civets with bamboo shoots and dried mushrooms. Or a dish called "The Dragon Battling the Tiger" will also warm you: stir-fry snake with civet cat and add a touch of red chili, according to a piece in *Time* magazine. A broth of gibbon bones cures epilepsy. Eat a tiger's eyes, and your eyes will become acute as a tiger's. The saiga antelope was exterminated in China because of the belief in the curative properties of its horns; China now imports saiga horns from Russia, about eighty tons in 1990 alone. The rare Yunnan golden monkey may also become extinct if its brains continue to be used for medicinal purposes. Pangolin scales are widely sought as medicine, and deer antlers in velvet bring such a high price that zoos lop and sell all antlers off their exhibit animals.

Between January and May 1987, over eight hundred pounds of musk, worth about fourteen million dollars at today's retail price, were smuggled out of China to Japan, the product of more than fifty-three thousand male musk deer. Most musk deer are snared, a capture method that kills animals of all ages and both sexes indiscriminately, but only males have musk glands. More than a hundred thousand musk deer died within a few months to treat Japan's ills. Although China has several small musk deer farms, the amount of musk they produce is trivial.

In 1987, restaurants in the city of Harbin consumed well over four thousand pounds of brown bear and black bear paws and seventeen hundred pounds of moose nose, according to the *China Daily.* In September 1990, a black bear was suddenly found roaming the city of Guangzhou. An investigation disclosed that officials of the local Wildlife Protection Society had smuggled that bear and eight others into town for lease to high-priced restaurants that used them to lure customers. A banquet of Chinese bear paws may cost a thousand dollars in Tokyo.

In Asia, bear gall bladders have long been considered such powerful cures for ulcers, fevers, burns, and other problems that bile fluid per ounce is now eighteen times more valuable than gold. In a 1991 report to WWF, Judy Mills and Christopher Servheen noted that Japan imported from China about fifteen hundred pounds of bear bile in 1989 alone. In addition to killing wild bears for their gall bladders, China has between five thousand and eight thousand Asiatic black bears and brown bears on bile farms. A tube is surgically implanted into a bear's gall bladder. At intervals the bear is confined to an iron squeeze cage so tight that it cannot move, and bile is drawn off into a bottle drop by drop. A bear can produce six to seven pounds of bile annually for the market.

Rhino horn is an ingredient of many traditional Asian medicines. It is said that potions of rhino horn reduce fevers and cure insomnia, epilepsy, high blood pressure, and other ailments. China uses about fourteen hundred pounds of horn each year—even though its own rhino population was so decimated by the eighth century that horn had to be imported even then. A member of the Convention on International Trade in Endangered Species (CITES) since 1981, China continued to trade in rhino horn even after a 1987 CITES ban on all internal trade. Under international pressure to register its stock of rhino horn, China finally succumbed in 1989 and reported at least ten tons. China together with Taiwan, Thailand, and South Korea are the only countries where illegal traffic in rhino horn remains highly lucrative: wholesale prices are around a thousand dollars per pound for African horn and ten thousand dollars for Asian horn. Esmond Bradley Martin, who has done more than anyone to elucidate the scope of the world's rhino horn traffic, discovered a sad case of cultural vandalism in China, as reported by him in the January 1991 issue of *Wildlife Conservation.* Craftsmen during the Ming and Qing dynasties carved intricate cups and bowls out of rhino horn. A cup made out of rhino would break, it was thought, when a drink poured into it contained poison. Medicine factories are now buying these art treasures from collectors and businessmen to grind into dust.

In 1982, the Ministry of Forestry planned to capture several of the country's last wild Siberian tigers to improve the genetic diversity and hence breeding potential of the zoo animals. To save the free-living tigers, the New York Zoological Society and WWF–Hong Kong coordinated a gift of eight tigers from United States zoos. The animals arrived in China in 1983. I thought no more about the matter, even when China established a captive breeding station for Siberian tigers in 1986 in Heilongjiang Province. There is such a world glut of this tiger in captivity that zoos elsewhere practice birth control. There were about fifty tigers at this breeding station in 1990, perhaps including some of the gift animals from the United States. I was at first unaware of the true reason for the breeding effort, run jointly by the Ministry of Forestry and a Chinese animal products company under the Ministry of Foreign Trade: its main purpose is to provide bones, whiskers, male sex organs, and other tiger parts for use in traditional medicines. Some one hundred and seventy-five pounds of tiger bones were made available in 1989. Said to be good for rheumatism, tiger bones sell for more than three hundred dollars per pound overseas; tiger penises sell for over one hundred dollars each. China has applied for a CITES permit to sell tiger products on the international market. A Chinese official told Esmond Bradley Martin on a visit to the breeding station, "If we don't get the permit, we'll just kill all the tigers."

China also attempts to earn much-needed foreign exchange by selling its wildlife to other countries. Several thousand mountain-dwelling blue sheep were shot annually in Qinghai Province from the late 1950s to the late 1980s, and the carcasses were then exported to Europe, especially to Germany, for the luxury meat market. There was no attempt at sustainable yield: blue sheep were simply decimated or eradicated in an area; deprived of their principal wild prey, snow leopard suffered too. In 1987, also in Qinghai, a German hunter was permitted to shoot white-lipped deer for a hefty fee and a gift of two jeeps. And now hunts for this handsome large deer are advertised at a fee of thirteen thousand dollars per animal. Yet white-lipped deer are strictly protected by Chinese law. In 1988, four American trophy hunters were permitted to kill Tibetan argali sheep in Gansu Province even though this majestic animal is internationally protected by CITES regulations.

Harvesting is a legitimate economic use of wildlife, but only if the animal is not endangered, the population is properly managed, and a significant portion of the income is applied to conservation and for the benefit of local people.

Indeed, in its rush for profits, China has neglected an alternative to destruction—the potential for wildlife tourism, both Chinese and foreign. For example, five crane species nest or make migratory stops at the Zhalong Reserve in Heilongjiang Province. The Book of Songs, China's earliest book of poetry, noted nearly three thousand years ago, "When the cranes sing in remote marshlands, their song can be heard in heaven." But this wild music may grow ever more faint. Villagers cut reeds for the manufacture of paper to such an extent that, deprived of nesting cover, the red-crowned crane could soon be in trouble.

Poyang Lake in Jiangxi Province was the site of one of the most exciting ornithological discoveries of the decade. A few Siberian cranes had occasionally wintered at the lake far from their Russian nesting grounds. During the 1980s the number of cranes there rose rapidly, perhaps due to the destruction of other wetlands, until during the winter of 1988–89 over two thousand six hundred were counted, almost the entire world population of these elegant white birds. Yet the *China Daily* reported that during the winter of 1984–85 one village cooperative alone killed some six hundred white storks, Siberian cranes, and whistling swans to make feather fans, and market hunters shot and poisoned about two hundred thousand wild fowl around the lake, despite a provincial ban against hunting. A public outcry finally brought some measure of protection. George Archibald of the International Crane Foundation has led enthusiastic tours of bird watchers to both Zhalong and Poyang, illustrating the kind of benign economic use that should be promoted to benefit wildlife as well as local people.

My litany of ecological woe must be balanced with an expression of praise for a government that realizes the negative effects of unrestrained development and has promoted conservation with considerable vigor and commitment. Indeed, no nation has accomplished so much in a mere decade and a half in spite of a shortage of trained personnel and money. There is a growing perception among officials that China's natural heritage exists not just for economic gain and that it must be preserved as a statement of the nation's vision for the future and identity with the past, as an integral part of the splendor of China's civilization.

China has had a long tradition of concern for its resources. As early as the Zhou Dynasty (1100-771 B.C.), the time and place for cutting timber were carefully regulated. After Liberation, the Third National People's Congress in 1957 resolved to establish forest reserves, and a 1962 directive from the State Council urged all provinces to "actively protect and reasonably utilize wildlife resources." A new Forestry Law as well as a Law on Environmental Protection were promulgated in the late 1970s. Private woodlands have been permitted since 1980, giving villagers an incentive to plant and care for trees on their own, and the conversion of grasslands to fields is being discouraged. Many appropriate laws have indeed been passed and pertinent orders and proclamations made, but such directives are often ignored at the local level, where quick cash receives priority over potential future yield.

No country has in recent years established as many nature reserves as China. Since the late 1970s, the Ministry of Forestry and the National Environmental Protection Agency, as well as the provincial governments, have designated new reserves with dizzying speed. From forty-four reserves in 1956, the figure rose to one hundred and four in 1981, three hundred and eighty three in 1986, and about six hundred in 1991. The goal is about eight hundred reserves with five percent of the country's territory. Several reserves are large. For example, in the Xinjiang Autonomous Region, a wild Bactrian camel reserve is four thousand square miles; the Taxkorgan Reserve which protects the last one hundred fifty or so Marco Polo sheep in the country is five thousand four hundred square miles; and the Arjin Shan Reserve at the northern edge of the Tibetan Plateau is more than seventeen thousand square miles.

In 1990 and 1991, the Chang Tang Reserve in northern Tibet was created to protect not just the unique upland fauna of wild yak, Tibetan antelope, and other species but also to conserve an intact ecosystem. About one hundred and fifteen thousand square miles in size, the reserve is larger than Colorado or the United Kingdom. Conservation initiatives are usually made in response to crises, after the wildlife has been decimated and the habitat almost destroyed. That China has made a commitment to protect such a huge area in today's crowded world is a conservation initiative of rare insight and importance.

It is relatively easy to designate reserves, but it is not easy to protect and manage them. China has serious problems in maintaining many of its reserves, as shown depressingly well in Wolong. However, a number of reserves have benefited from various conservation initiatives. For example, farmers who lived within the Tangjiahe Reserve have been resettled. In Yunnan's Wuliangshan Reserve, a team of twenty trained guards protect the black-crested gibbons and other wildlife. Most important, people around the reserve are being encouraged to plant fruit trees, tea, and other cash crops rather than depend for income on hunting and wood-cutting within the reserve.

Concerned with restoring that which has been squandered, the government began a reintroduction program for Mongolian wild horses. Exterminated in the wild during the 1950s, this horse may within a decade or two reoccupy parts of its former home. With sixteen zoo-bred horses obtained from San Diego, Berlin, and Munich, a herd was established in a thousand-acre walled enclosure in the desert of the Junggar basin of northern Xinjiang. Once herd size grows to eighty, some animals may be liberated. There is another captive herd, destined for future release, in Gansu Province.

Although the Père David deer has been extinct in the wild for perhaps a thousand years, some survived in the imperial hunting park near Beijing, a walled tract of woodland about one hundred and fifty square miles in size. These deer were killed and eaten by soldiers and villagers in 1900 during a rebellion led by the Society of Righteous and Harmonious Fist, the so-called Boxers. Fortunately a number of deer had by then been taken to Europe, and these the duke of Bedford brought to England and bred on his estate, Woburn Abbey. In November 1985, through the efforts of Maria Boyd and her Chinese colleagues, twenty-two Père David deer came home. The herd was settled into its old haunts, a walled four hundred and forty acres on the former site of the imperial park. More than a hundred deer now live in the enclosure. A second captive herd has been established in Liaoning Province, and a third is planned for Hubei Province.

The construction of a breeding facility for pandas in Wolong was part of a trend by the Chinese government to protect an endangered species by raising it in captivity. It disconcerts me that, however noble its intentions, China spends millions of conservation dollars on building facilities for pandas, tigers, and other animals when, more than anything else, the money is needed to protect species in their natural habitats.

Certain hardy animals such as the Chinese alligators and Mongolian wild horses have benefited greatly from captive propagation. They breed readily under confinement, and can with only modest difficulties be returned to the wild. However, delicate species, among them François' langur monkeys and golden monkeys, are also the focus of breeding efforts. Three golden monkey roundups, during which many villagers encircled a forest and drove the frantic monkeys into a stockade, were carried out in Shaanxi, one as recently as 1979, according to Tan Bangjie of the Beijing Zoo. Of about three hundred and seventy-five monkeys caught, two-thirds were dead within two weeks. The question should be asked, captive breeding for what? It is exceedingly difficult to introduce captive-born primates into the wild—the main justification for any breeding program—and besides their native habitat may vanish unless it is protected. Today nearly four thousand black-necked cranes still exist on the Tibetan Plateau. To nest successfully, they need protection from disturbance by livestock, people, and dogs in their breeding marshes. Instead of solving this relatively simple problem, the government has discussed the idea of an expensive captive program. As the Chinese would say, "There is money for coffins but not for medicine."

Captive breeding should be a last resort, used only after genuine efforts to maintain a species in the wild are clearly failing. Such efforts have not yet been made. Often forgotten is the fact that the panda, tiger, and others are merely symbols, so-called flagship species, of the natural environment in which they occur. The real national treasure of China or any country is the habitat with all its animals and plants; it provides watershed protection, recreation, a genetic storehouse of unique species, and other resources upon which the lives of people will depend.

Kylin are dainty creatures; they have the body of a deer, a long tail, and a single straight horn. They make a bell-like sound. When moving about, they will not tread on any living organism, not even on grass. They drink only pure water. In the fifth century B.C., Confucius was told of a kylin, but by the time he arrived at the site a hunter had killed it. To the Chinese, the kylin embodies justice and virtue. To see one is exceptionally auspicious, for such a sight means that the nation's future will be prosperous. If China will but treat the fragile beauty of its land with greater compassion, the kylin may appear once again.

APPENDIX B

The Panda Is a Panda

But I do beguile the thing I am by seeming otherwise.

William Shakespeare, *Othello*

Is the giant panda a bear? Are the red and giant pandas closely related? These two questions have been debated for over a century. Anatomists, behaviorists, paleontologists, and molecular biologists have led the fascinating inquiry into the evolutionary relationships of these species with ingenuity and persistence, yet they continue to derive different conclusions on the basis of different evidence, and they still pursue the elusive answers. Their quest well illustrates the logic and method of scientific inquiry, the search for a resolution that satisfies the intellect. I have no emotional investment in the outcome, such research being peripheral to my interests, but I find the unraveling of the riddle intriguing.

Red pandas and giant pandas are inextricably linked. They share not only a name but also many physical similarities. Skull, teeth, and forepaws in each are similar, evolved to process bamboo. They even grip bamboo in much the same manner, except that the red panda lacks the functional sixth digit or pseudothumb so useful to the giant panda for manipulating stems. No one questions that the two pandas resemble each other. But are they actually related? This has been a subject of scientific controversy since Père David discovered the giant panda in 1869. He gave it the generic name *Ursus* on the assumption that the animal was a bear. The following year, Alphonse Milne-Edwards looked at the same skeletal material and decided that the giant panda was not a bear but allied to the raccoons. Since then the two pandas have been bounced around in an ever more esoteric and technical manner from one taxonomic home to another.

The issue is basically simple. Some biologists looked at the two pandas and decided they were not closely related, that their physical similarities had evolved because of the same food habits and life-style. They placed the red panda into the raccoon family, the Procyonidae, and the giant panda into the bear family, the Ursidae. Other biologists looked at the same evidence and came away convinced that the two *were* relatives, belonging to the same branch of the evolutionary tree. They placed the pandas either into a separate family or tucked them in with the raccoons. Each school of thought could point to specific features to reinforce its claims. The bear school, for example, stressed the giant panda's bearlike body proportions, but the raccoon school countered that the giant panda's skeleton was unusually heavy-boned, giving "the impression of being the skeleton of a 'fake' bear," as Ramona and Desmond Morris wrote. The giant panda certainly looks like a bear, but is it a bear pure and simple? Or is it a raccoonlike animal that resembles a bear because it has grown large, with a heavy body that requires special modifications such as stout legs to support it?

In some ways, the long-running issue is trivial, an illustration of a scientific discomfort with uncertainty and a penchant for putting everything tidily in place. If the giant panda is a bear, it is a highly aberrant one. If the giant panda is a raccoon, it can look around at a peculiar hodgepodge of family members: the long-nosed coati, the prehensile-tailed kinkajou, the conservative ringtail, a living fossil little changed from it Oligocene ancestor thirty million years ago, and the nimble-fingered raccoon.

However, the controversy also poses a fundamental scientific problem—what features are important and significant when classifying an animal? Classification was difficult enough when only physical characteristics were considered, but with the advent of molecular biology more, and often conflicting, lines of evidence had to be evaluated. Surprisingly, the puzzle of the panda's origin has seeped into public awareness, and the problem is viewed with some of the same fascination as the disappearance of dinosaurs. When giving a lecture, I am often asked at the end whether the giant panda is a bear or raccoon. To keep my reply brief, I usually answer, "The panda is a panda."

This view places me in a distinct minority at a time when the bear proponents are the most vocal. Take a scientific review on the subject published in 1986 in the journal *Nature.* The author, clearly impatient with the whole controversy, noted that some still view the two pandas as "each other's closest relatives. It would seem that this suggestion is clearly refuted by the near-unanimous and highly diverse evidence . . . ," and he hoped that the "overwhelming evidence for the bear relationship of the giant panda would end the argument. . . . Only some of the behaviour students were not yet persuaded." When "overwhelming" evidence is invoked, beware. Is the article a polemic promoting a dogma rather than a careful examination of the facts? In science, an issue that has been truly resolved is absorbed and disseminated quietly without need to make the "near-unanimous" unanimous, to prod heretics into becoming true believers.

Am I reluctant to become a bear proponent because of mental inertia, am I unable to divest myself of an outdated notion, or is there reasonable doubt about the majority view? I am admittedly delighted that in spite of its being dissected, observed, measured, and subjected to a host of advanced molecular techniques, the giant panda still cannot be neatly categorized. Just as I hope that there is a yeti but that it will never be found, so I would like the panda to retain this minor mystery. Still, there is intellectual pleasure in trying to solve such a puzzle.

Molecular studies have in recent years provided important insights into the classification of pandas. Since protein molecules are an integral part of DNA, they closely reflect the hereditary and therefore the evolutionary history of an animal. Proteins are composed of different amino acids. The number of differences between the amino acids of any two proteins should, it was reasoned, be proportional to the time elapsed since they diverged from a common ancestor. With molecular evolution supposedly freed from environmental influences, it might be regular enough to be used as a molecular clock, one that provides precise information about the amount of genetic change that has occurred over the past few million years. In one such study, published in 1976, Vincent Sarich used immunological techniques to compare two blood proteins from giant pandas, red pandas, bears, and raccoons. In this method the proteins are injected into rabbits, producing antibodies, which then react strongly against the proteins for which they were prepared and progressively more weakly against proteins in species more distantly related. He concluded that "the association of the Giant Panda and other bears is clear and unequivocal. . . . The one rather unexpected result there is the fact that the Lesser Panda, *Ailurus,* does not group with the other procyonids," that it seems to have started a separate lineage before the giant panda and bears.

In 1985, Stephen O'Brien and his coworkers published an article in *Nature* entitled: "A molecular solution to the riddle of the giant panda's phylogeny." Using gel electrophoresis, a technique by which proteins can be sorted by electric charge and size, proteins from pandas, bears, and members of the raccoon family were compared. To check their results, the investigators conducted DNA hybridization tests. This method compares the DNA of each species, the actual hereditary material, rather than just the proteins. Radioactively tagged DNA from one species is hybridized with the DNA of another and the stability of the union measured. The study concluded that "the lesser panda diverged from New World procyonids at approximately the same time as their departure from ursids, while ancestors of the giant panda split from the ursid lineage much later, just before the radiation which led to modern bears." This split is said to have occurred as long as fifteen to twenty-five million years ago. In other words, the giant panda is a bear, the red panda a raccoon. Yet within a year after this study, another investigation into a blood protein showed that the two pandas are more alike than are either giant pandas and bears or red pandas and raccoons. Which blood protein should be used as a taxonomic character: albumin, which makes the giant panda a bear, or hemoglobin, which does not? The molecular clock may at times not be as precise as suggested. Different proteins in an organism may mutate at different rates, and natural selection may affect some proteins, such as hemoglobin, more than others.

David Goldman, Rathna Giri, and Stephen O'Brien returned to the fray in 1989 with a molecular study of all seven bear species, the giant pandas, the red pandas, and the raccoon. Published in the journal *Evolution,* the study was "based on the extent of electrophoretic variation of 289 radiolabeled fibroblast proteins resolved by two-dimensional gel electrophoresis and among 44 isozyme loci resolved by one-dimensional electrophoresis." I am too ignorant of molecular biology to understand such studies, much less to comment critically on techniques and analyses. The results generally confirmed Stephen O'Brien's 1985 report. The progenitors of raccoons and bears split during the Oligocene, and within ten million years the red panda lineage diverged from the raccoon lineage. In the Miocene, there were three major radiations among the bears, the earliest line leading to the giant panda, the second to the South American spectacled bear, and the last to the six other bear species. If the authors expected to settle the panda debate with this study, it must have been a forlorn hope. Indeed, given the history of molecular studies, the next one would most likely contradict their findings. And so it did.

In 1991, Zhang Yaping and Shi Liming published in *Nature* the results of a detailed analysis of mitochondrial DNA from the two pandas, Asiatic black bear, and sun bear. Their conclusion: the giant panda is more closely related to the red panda than to the bears. And they noted that, "there are 1,000–10,000 copies of mitochondrial DNA in every cell; the selection pressure on this DNA is very low. So the similarities between the two pandas . . . may not be the result of convergent evolution."

The contradictory results from molecular studies teach an important scientific lesson. Slick and ultramodern techniques may not always be enough to elucidate the incomprehensible. Morphology, paleontology, and natural history continue to have an important role in unraveling the evolution of the two pandas.

Fossils often provide insights into the past of a species, but there are giant gaps in the fossil record, especially of pandas. The raccoon family, an early branch of the dog family, the Canidae, evolved in North America and spread across Asia to Europe where, during the Miocene, twenty million years ago, the raccoonlike *Sivanasua* was found. The bears, also a branch of the Canidae, appeared in the early Miocene in the form of *Ursavus,* a bearlike creature as large as a medium-sized dog. The first definite panda, *Parailurus,* a small animal resembling the red panda, occurred in the early Pliocene about twelve million years ago in southern Europe and North America; it persisted into Europe's last ice age, where it was quite common in temperate forests. Some researchers think that a small, bearlike animal of the *Ursavus* lineage named *Agriarctos,* dating from the mid-Miocene, was the ancestor of the giant panda. Support for this supposition came in 1989 when Qiu Zhanxiang and his colleagues described a new fossil from the late Miocene in Yunnan with teeth that resemble those of the giant panda but also share characters with the ancestral forms of bears. The animal, given the name *Ailurarctos lufengesis,* was less than half the size of today's giant panda.

The giant panda itself appeared suddenly during the late Pliocene or early Pleistocene, perhaps no more than two to three million years ago. Panda fossils have been found in Burma, Vietnam, and particularly in eastern China, as far north as Beijing, where they appear so often with the Pleistocene elephant *Stegadon* that the two species are used to designate a distinct fossil fauna. The pandas of the early Pleistocene were about half the size of today's giant panda and are considered a separate species, *Ailuropoda microta.*

New species originate mainly when a small segment of an ancestral population becomes isolated, and the animals change form and behavior through natural selection until a population with new characteristics is well established. This process may be so rapid that no recognizable intermediate forms or missing links are found in the fossil record. Animals can apparently evolve quickly, through major chromosomal rearrangements, as well as more slowly through mutations of single genes. Then, having settled in, the new species may remain unaltered, except for slight modifications, for millions of years. The giant panda is known only in its existing form, apparently not an ancient relic, as is often claimed, but a relative newcomer. Its previous incarnations still remain uncertain.

Can the behavior of the giant panda provide clues to the animal's evolutionary relationships? This approach has problems. Species living in similar habitats may evolve similar societies and similar physical appearances, which in turn may result in similar behavior without there being a close relationship. When comparing species, one first has to decide which aspects of behavior have been strongly influenced by ecological conditions. For instance, the amount, quality, and distribution of food effects an animal's movements, activity cycles, and social structure. Consequently two separate populations, even of the same species, may behave differently. However, certain kinds of behavior such as scent marking and vocalizing can function well under a wide variety of conditions, and therefore they may be less influenced by ecological pressures.

The giant panda produces a surprising mix of sounds, some of which it shares with bears, some with the red panda, and some with both. For instance, the giant panda's chomping, in which the animal clacks its teeth and smacks its lips when anxious, is found in bears and, in a modified form, also in red panda and even coati. The giant panda's plaintive honk, which denotes light distress, is similar to the grunts made by bears and several procyonids, but in these animals the calling has a different function, that of a contact call between mother and young. More exclusive is the moan, a highly variable call ranging from hoots and whiney groans to long-drawn-out moans. Only giant panda and bear share this warning signal. Particularly noteworthy is the giant panda's goat-like bleat, a friendly call that provides animals with reassurance on meeting. As Gustav Peters of the Alexander Koenig Museum in Bonn has noted, this bleat has its equivalent in the twitters and chitters of red panda and all the procyonids, but nothing similar occurs among the bears. One would not expect such a high-pitched vocalization from an animal the size of a giant panda; in fact, several of the giant panda's calls are surprisingly high-pitched for a large carnivore. Nursing bear cubs produce a most peculiar call, a continuous keckering, harsh and rapid almost like a loud purr. Although the function of this vocalization is unknown, it obviously conveys something important to the mother, possibly signaling her to lie still and release milk. The giant panda lacks this call. If the giant panda is merely a bear, I find it difficult to understand why the tiny young would not have such an emphatic vocalization.

Marking behavior naturally depends on the type of gland available to a species and on the location of the gland. The giant panda claws trees, urinates, and rubs its glandular anal area on objects. Bears lack a specific gland, although they stand on their hindlegs and rub shoulders, neck, and head against tree trunks, and they bite and claw bark, leaving behind their general body odor. The red panda straddles stumps and other protuberances and deposits scent from its anal glands with circular rubbing motions, using actions closely resembling those of the giant panda; male red pandas also squirt urine. Miles Roberts of the National Zoo in Washington, D.C., found that red pandas also have several small pores on their palms that secrete a clear fluid, a means of leaving special information behind when walking. Most procyonids scent mark with urine or anal glands or with both, behavior similar to that of pandas though the details differ.

Giant pandas and bears also resemble one another in the extremely small size of their newborns. While black bear mothers are two hundred and fifty times heavier than newborns, a giant panda mother is nine hundred times heavier. (By contrast, a raccoon mother is only fifty-five times heavier.) Why is the giant panda newborn so much tinier than the others? Reproduction is influenced by ecological conditions, especially by the amount of high-quality food available to mother and young at various seasons. Bears, like giant pandas, have delayed implantation. In temperate climates bears mate and young are conceived in about June but the fertilized egg does not attach itself to the uterine wall and growth of the fetus does not begin until about sixty days before cubs are born in January or February, while the mother is in hibernation. If cubs were large and vigorous at birth, their milk demands on the lethargic mother might be excessive, too great an energy drain on the fat deposits that must maintain her until the end of hibernation in spring. Since the giant panda does not hibernate, different selection pressures must have produced the tiny newborn. The fact that bears and giant pandas both have small young at birth does not imply that they are kin. I wondered, then, why a giant panda newborn was so extraordinarily small.

Looking at red pandas, I found some clues. The red panda also has delayed implantation (something not found in procyonids), and its gestation period is an average of one hundred and thirty-one days, about the same as that of a giant panda. The daily weight gain of a giant panda fetus is less than half that of bears and more like that of the red panda as well as raccoons. A red panda mother usually has one or two young and their birth weight is about four ounces, the same as the birth weight of the giant panda. The red panda female keeps her young hidden in a tree cavity where they develop slowly, emerging only at the age of three months, a denning period longer than that of raccoons and coatis. All these facts show that reproduction in a giant panda is more typical of a small mammal like a red panda than of a bear. There are of course some size adjustments. For example, the long period of care required by the young giant panda extends the mother's lactation period into the next mating season, preventing her from reproducing annually. By contrast, red panda young become independent at eight instead of eighteen or more months, and the female can reproduce each year. I deduce from all this that the giant panda has retained many reproductive features of a small pandalike ancestor, merely adapting some of the historical vestiges to meet current circumstances.

Scattered pieces of evidence point to a definite relationship between giant panda and red panda: the specialized structure of skull, teeth, and forepaws; various aspects of reproduction; certain vocalizations; and scent-marking behavior. If the two species are unrelated, we must accept a remarkable amount of convergence, more than is justified by the evidence. When did their paths diverge? Most likely the giant pandas and red pandas had a common ancestor in the Miocene. Where should the two pandas then be placed, with the bear or with the raccoon family? Even though the giant panda is most closely related to the bears, I think that it is not just a bear. Even a small amount of genetic difference between two species may have a profound influence on appearance and behavior. Chimpanzees and humans may share as much as ninety-nine percent of their genetic material. We should no doubt embrace the chimpanzee as a family member. But is the chimpanzee human? The giant panda and red panda offer several choices. Should the giant panda be with the bears or in a separate family but the red panda with the raccoons? Should each panda have a separate family? Or should the two pandas share a family, the Ailuridae? I favor the last alternative. Science will overcome this paradox and perversity of evolution and ultimately assign each panda a final taxonomic home.

But as yet this game of taxonomic Ping-Pong has no winners. The search for a more complete answer continues "with the bear proponents and the raccoon adherents and the middle-of-the-road group advancing their several arguments with the clearest logic, while meantime the giant panda lives serenely in the mountains of Szechuan with never a thought about the zoological controversies he is causing by just being himself." Edwin Colbert wrote these words in 1938. The giant panda still pseudothumbs its nose at us.

APPENDIX C

Winter Birds Observed around Wuyipeng, Wolong Natural Reserve, 8,000–10,000 feet

(Over two hundred and thirty bird species have been recorded in Wolong, but most are migrants or live at low elevations. During winter, I found only these thirty-five resident species around our camp.)

Common Name	Latin Name
Severtzov's grouse	Bonasia sewerzowi
Blood pheasant	Ithaginis cruentus
Temminck's tragopan	Tragopan temminckii
Koklass pheasant	Pucrasia macrolopha
Chinese monal pheasant	Lophophorus lhuysii
Tawny owl	Strix aluco
White-backed woodpecker	Picoides leucotus
Darjeeling woodpecker	Picoides darjellensis
Great spotted woodpecker	Picoides major
White-throated dipper	Cinclus cinclus
Northern wren	Troglodytes troglodytes
White-crowned forktail	Enicurus leschenaulti
Little forktail	Enicurus scouleri
Streak-throated fulvetta	Alcippe cinereiceps
White collared yuhina	Yuhina diademata
Pygmy wren babbler	Pnoepyga pusilla
Giant laughingthrush	Garrulax maximus
Barred laughingthrush	Garrulax lunulatus

Elliot's laughingthrush	Garrulax elliotii
Fulvous parrotbill	Paradoxornis fulvifrons
Great parrotbill	Conostoma oemodium
Gray-crested tit	Parus dichrous
Coal tit	Parus ater
Rufous-vented tit	Parus rubidiventris
Père David's tit	Parus davidi
Green-backed tit	Parus monticolus
Great tit	Parus major
Eurasian nuthatch	Sitta europaea
Common treecreeper	Certhia familiaris
Japanese white-eye	Zosterops japonica
Gray-headed bullfinch	Pyrrhula erythaca
Collared grosbeak	Mycerobas affinis
Eurasian nutcracker	Nucifraga caryocatactes
Gold-billed magpie	Urocissa erythorhynchus
Large-billed crow	Corvus macrorhynchos

APPENDIX D

Excerpts from the 1989 "National Conservation Management Plan for the Giant Panda and its Habitat"

(These excerpts summarize the plan's main recommendations for panda management. Copies of the 157-page plan can be be obtained from the World Wide Fund for Nature, Gland, Switzerland.)

Discussion of Management Options

Reduction of Human Activities in the Panda Habitat

The greatest threat to the giant panda is people. The overall decline of the panda is not a story of flowering bamboo, it is a story of a species pushed out of its habitat by human expansion. . . .

Suitable panda habitat can be rather summarised as large interconnecting areas of not too steep terrain with moderate tree cover and extensive undergrowth of at least one (preferably more) species of bamboo, undisturbed by human activity, dogs or domestic livestock and with adequate stream-flow to provide year-round water supply.

The advances of human farming, logging, hunting, and grazing of livestock over the last 150 years have resulted in considerable degradation and fragmentation of original panda habitat. To save the panda it is necessary to reverse the process and reduce human activities in panda habitat by active management.

Such management can be applied in two main ways:

1. establishing a network of panda reserves where the conservation of pandas and other wildlife is the primary management objective and where human uses should be limited to those that are compatible with the objective; and
2. improving panda survival in suitable habitat outside the reserve system where pandas still occur, by limiting human activities and modifying forestry operational practice in such a way that pandas can continue to survive there.

In nature reserves, the normal regulations prohibiting human settlement, cutting of forest, collection of medicinal plants, grazing of domestic animals and hunting should be strictly applied to all citizens, including minorities. In panda habitat outside of reserves some new protective activities should also be applied:

—no hunting, or burning of vegetation permitted
—forestry operations to be modified so as to limit damage to pandas
—active bamboo restocking where necessary
—reafforestation only with local species
—protection of reestablishment of migration corridors across all roads and where critical exchange routes are identified
—regular monitoring of the panda population and bamboo
—halt those activities such as logging that prove to be harmful to the survival of the giant panda

Removal of Human Settlements

The presence of people inside nature reserves causes an impossible conflict of interests. People can be excluded by selecting or redrawing reserve boundaries to exclude human settlements or where this is impossible to actually move and resettle human communities. A start has been made in reversing the trend of increasing human encroachment by relocating a total of 300 persons or 60 households from the Tangjiahe Natural reserve in Qingchuan County of Sichuan Province. This was achieved by getting local government to undertake the resettlement of these people in neighbouring communes outside the reserve and by payment of a negotiated sum of money by the Forestry Ministry to the local government to cover this expense and to ensure adequate compensation to the villagers themselves for loss of resources such as orchards and walnut trees left behind in the reserve. . . .

Modification of Forestry Operations

China faces a huge wood shortage. It is inevitable that the country cannot afford to leave all the timber in panda habitat uncut. However it is possible to considerably reduce the damage to panda habitat by applying the following principles:

1. No new timber units will be established in the giant panda habitat.
2. Existing timber units will have their cutting area gradually reduced and will have to modify their cutting methods.
3. Apply selective cutting techniques with small scale operations so that a reasonable canopy (not less than 30%) is left to shelter bamboo and the forest can reseed itself naturally. All large hollow trees should be left as seed sources and potential panda nursery dens.
4. Cut the forest in a pattern that preserves routes and quiet habitat for pandas to emigrate during the time of logging disturbance.

Control of Poaching

Despite heavy penalties for poachers found hunting inside reserves, or for anyone killing a panda anywhere, poaching is still a serious threat to the giant panda. . . . Poaching can only be controlled by more rigorous patrolling, bringing more poachers to trial, tightening up the legal loopholes and closing down the trade outlets for animal products. . . .

Rehabilitation of Habitat

Where natural habitat has been destroyed, but the distance between occupied habitat is not great, it is worth applying habitat rehabilitation to reconnect panda populations. Such rehabilitation will generally involve replanting suitable species of bamboo and forest cover. A number of constraints must be considered:

—some bamboos inhibit the growth of tree seedlings so the establishment of forest stock may have to precede bamboo replanting;
—rehabilitation should focus on habitats known to be preferred by pandas with relatively level terrain and proximity to water;
—rehabilitation should recognise the need for a diverse bamboo ecosystem and allow replanting of trees and bamboo species in appropriate elevation ranges; and
—tree planting must be done less densely than for production forests to allow a good understorey to develop. A natural mix of local species should be used.

Introduction of bamboo is a laborious task, especially where large areas have to be covered. It is important to select the most suitable species, preferably those that formerly dominated that particular locality. . . . *On no account should exotic bamboos be introduced* into the ecosystem. There are several instances in other countries where that has proved environmentally disastrous. . . .

Management of Bamboo Habitat

Periodic mass flowering followed by death of adult plants is almost universal among temperate bamboo species and is not a recent phenomenon. We must draw two conclusions—firstly that such phenology is adaptive and therefore important for the survival and health of the bamboo, and secondly that given a normal range of altitudinally distinct bamboo species, the panda is well enough adapted to cope with such events even if this is naturally a period of increased mortality among pandas. We must rid ourselves of the idea that human interference in the form of broadening the species diversity of bamboo and trying to artificially stimulate or delay the flowering process (to reduce the severity of food loss after flowering events) is a necessary or desirable form of management. It is both unnecessary and carries inherent risks for the local bamboo ecology and ultimately the panda. . . .

Several arguments suggest that periodic local die-off of bamboo is probably essential for the survival of bamboo, essential to permit the forest regeneration cycle and important for pandas in applying selection and promoting outbreeding (through emigration) between weakly selected, inbred, drifting genepools of giant pandas. We are dealing with a coevolutionary tangle of considerable complexity and cycles of balance that must be seen over a 50–100 year timespan not merely a couple of decades. It would be risky to meddle with relationships we do not fully understand. . . .

Extension of the Panda Reserve System

It is clear that most of the existing panda reserves are too small to contain viable panda populations and it is therefore necessary to enlarge and strengthen the existing reserve system to link the reserves, where possible, by giving partial protection to intervening panda habitat. . . .

Achieving Outbreeding between Panda Populations

Outbreeding can be ensured by preserving remaining corridors of suitable panda habitat linking the different population units. . . . However, where natural links are irrevocably broken, we must apply active management to stimulate natural levels of outbreeding. Genetic analysis of sub-populations would reveal population differences and indicate optimal crosses or translocations. Outbreeding can be achieved by 3 methods:

1. Reestablishing forest and bamboo corridors to link separate population units.
2. Translocation of animals between different populations.
3. Introduction of captive born young into the existing wild population. . . .

Maintaining a Captive Population

. . . The number of pandas already in captivity is about 80 animals and this is quite enough to form a self-sustaining breeding population. . . . A successful captive breeding programme would be a useful tool in trying to save the species.

However, it is dangerous to capture more wild individuals for this purpose. We cannot claim to have saved the giant panda if all we end up with is a viable captive population. The real goal of conservation is to save the species in the wild by means of habitat conservation and maintenance of genetic variation. Captive breeding is only useful for species conservation if it helps to achieve that end and this is only possible if we find ways to release captive born pandas back into the wild. As yet success of captive breeding is severely hampered by inadequate skill in managing and breeding captive pandas and inadequate facilities at many of the institutions holding captive pandas, poor cooperation between the different institutions holding pandas, most of which are under the control of the Ministry of Urban Construction and not the Ministry of Forestry, makes matters worse. At present the breeding rate in captivity is still very low, mortality of captive born pandas is too high and the captive population of pandas continues to be a drain on the wild population rather than a booster to the wild population. It is proposed to put a ban on any more wild-caught pandas reaching zoos. Closing this outlet for restocking will force the zoos to try harder and cooperate better in promoting captive breeding. . . .

Release of Captive-born Pandas into the Wild

. . . The ultimate aim of breeding pandas in captivity should be to acquire animals for use in the restocking of wild populations to help populations to expand into formerly occupied or newly rehabilitated range, to help populations to recover from crashes after bamboo flowering events and to enhance genetic diversity in wild populations. Suitable methods need to be developed to achieve this. So far little work has been tried with the release of captive pandas but there have been a few successful releases of wild-born pandas.

All pandas readily recognise bamboo as food in captivity and seem to instinctively know how to deal with it. Having a simple lifestyle, browsing on a common ground plant, it is unlikely that pandas will require much special learning before release. It would be dangerous to release naive young pandas into areas where there are leopard or hunting dogs or aggressive resident pandas but young adults should be able to cope for themselves against such predators.

. . . It is necessary for some bold decisions to be made and some trial animals to be released into areas where there is plenty of bamboo and they can find food easily.

APPENDIX E

Excerpts from the Position Statements on Exhibition Loans of Giant Pandas

(Three organizations issued position statements on panda loans in 1988. I quote the relevant portions of these statements to show how each organization viewed the situation before urging a moratorium on all loans in 1990.)

WWF—World Wide Fund for Nature, January 1988

In recent years there have been an increasing number of exhibition loans of pandas to captive collections outside China. WWF recognises that these loans attract considerable public interest. The funds generated may benefit ex-situ conservation in China.

However, WWF has become increasingly concerned that since these loans subtract potential breeding animals from the captive populations, they do not form a useful part of an integrated breeding programme. WWF has raised this issue with the relevant organisations in China on several occasions. It has been suggested that the loans be restricted to animals which are either too old or too young to breed. However it is difficult to specify an age above which pandas are too old for breeding, and it would appear unwise to subject young animals, which might reproduce in the future, to the risks of international travel.

Therefore, WWF urges zoos outside China and the Chinese authorities to cease their involvement in exhibition loans of giant pandas, once the loans underway in Calgary, Canada, and Sydney and Melbourne in Australia are completed. WWF will not associate itself with any further loans in the future. WWF also urges the Chinese authorities to halt the use of giant pandas in circuses.

IUCN—The World Conservation Union, February 1988

RECOGNIZING that the giant panda, *Ailuropoda melanoleuca,* is endemic and now unique to the People's Republic of China, and has been a species of great interest to biological science for over a century;

APPRECIATING that the Government of the People's Republic of China has devoted much attention and considerable financial resources to the conservation of the giant panda;
NOTING that the giant panda is a species that is unusually well-known to the peoples of the world, and, through use of its image by WWF, has come to symbolize the universal need for wildlife conservation to governments and peoples of the world;
UNDERSTANDING that there may be fewer than 1000 of these animals in the wild and 100 in captivity, critically low numbers for survival of the species;
RESOLVING that every effort should be made to keep the giant panda from extinction;
The General Assembly of the IUCN, at its 17th Session in San José, Costa Rica, 1–10 February 1988:
1. COMMENDS the Government of the People's Republic of China for wildlife laws and regulations that protect the giant panda and for establishment of reserves to maintain the giant panda's natural environment;
2. STRONGLY ENCOURAGES the Government of the People's Republic of China to respond to increasing worldwide concern for the survival of the giant panda by adopting a comprehensive conservation plan for the species and fully implementing that plan as quickly as possible, including full utilization of the scientific, technical and educational resources of institutes and universities of the People's Republic of China.
3. SPECIFICALLY RECOMMENDS adoption of a long-term strategic and tactical conservation plan for the giant panda;
4. CALLS UPON all agencies and individuals in the People's Republic of China to cooperate in such a long-term conservation plan, preferably under a single directorate so that actions can be coordinated and controlled to best effect in securing the giant panda in its natural environment, in preventing poaching and habitat destruction, and in ensuring that any animals held in captivity contribute to the maintenance and perpetuation of the giant panda population as a whole;
5. URGES authorities of the People's Republic of China to provide for the early rehabilitation to the wild of pandas rescued from temporarily or permanently degraded habitats;
6. FURTHER URGES that all organizations and institutions now holding giant pandas cooperate fully to incorporate these animals in an international program for captive propagation and that the free interchange of captive specimens for this purpose be facilitated by all governmental and intergovernmental agencies concerned;
7. RECOMMENDS that the effects of other activities involving giant pandas that may not directly contribute to the conservation of the species, especially temporary exhibitions, be investigated by the appropriate authorities and scientists of the People's Republic of China in consultation with international conservation groups and the zoo associations of several countries, and that until the issues involved in these matters are resolved that careful consideration should be given before further loan agreements are made.
8. FURTHER RECOMMENDS that, in consideration of the low population num-

bers of the species, scientists and governmental authorities of the People's Republic of China specifically consider the proposition that any exhibitions of giant pandas should only be adjunct to and completely compatible with an international captive breeding program for the species, and that such exhibits should be designed for the education of people about the biology and conservation of the giant panda; and that any revenues specially generated from exhibitions should only be devoted to support specific aspects of the conservation plan for the species and accounted for in keeping with this goal;

9. ACKNOWLEDGES AND APPLAUDS the efforts of WWF, Wildlife Conservation International–New York Zoological Society, Zoological Society of London, and others in assisting in the conservation of the giant panda by lending expertise and giving financial support to colleagues and agencies of the People's Republic of China; and

10. AGREES to stand ready, within available resources, with its member organizations and WWF to assist responsible authorities in the People's Republic of China in the development and implementation of a comprehensive conservation plan for the giant panda, including design and execution of needed research, advice on park and reserve management, coordination of captive breeding programs, establishment and management of an international studbook, and development of popular education projects.

AAZPA—American Association of Zoological Parks and Aquariums, February 1988

While there continues to be a high level of interest in short-term loans, it is important that AAZPA members seeking loans do so with the welfare of the giant panda uppermost in mind. . . .

The Short-Term Loan Program

We must assure that short-term loans do not prove detrimental to the panda's future by removing pandas from possible captive breeding programs in China, or by preventing their reintroduction to nature, or by encouraging the removal of pandas from the wild.

A contracting zoo and their Chinese counterparts understand that the zoo's contribution to giant panda conservation would only be allocated through the agency of a panda conservation committee endorsed by the AAZPA.

AAZPA members should only accept animals for short-term loans which are adult specimens physiologically incapable of reproduction, preferably adult males and preferably animals that are captive born.

Transfer of animals should be done as infrequently and as quickly as possible.

Giant pandas must be accompanied during shipment by persons knowledgeable and capable of caring for the animals at all times.

Proper facilities must be available in advance of the arrival of animals.

Afterword

The Last Panda went to press late in 1992, but the panda saga has continued, marked by the usual combination of noble effort and pungent disregard for the animal. The rent-a-panda program in its traditional form ended in 1992. An international moratorium on such loans was in effect, the Columbus Zoo had been temporarily suspended by the American Association of Zoological Parks and Aquariums (AAZPA, now the AZA) for breaking the moratorium by importing a pair of pandas, and the United States Fish and Wildlife Service (FWS) refused to consider further import permits until it had reviewed its panda policy once again. I assumed that the zoo community would move forward with a collective vision and speak with a unified voice to work with China, developing an integrated conservation strategy for the panda in the wild and in captivity. I should have been devoid of such illusions and naiveté. Yet hope never dies.

Despite the moratorium and the suspension of the Columbus Zoo, the San Diego Zoo initiated secret negotiations with China for a pair of pandas. In a new reincarnation of the rent-a-panda program, the zoo asked for a three-year breeding loan with the hope that the pair would produce an offspring. With the pandas' future increasingly precarious, the zoo explained that it needed this pair to help save the species. However, to what extent a potential for profit may enhance a moral stance is an interesting point for debate. When San Diego first exhibited pandas in the late 1980s, the zoo's attendance in certain months was the highest in its history. A baby panda would guarantee the zoo a fortune. China agreed to the loan for the low price of one million dollars a year; in the event of a cub, the Chinese would receive an extra hundred thousand dollars a month up to a maximum of six hundred thousand dollars. So confident was the zoo of circumventing the moratorium and pressuring the FWS to issue an import permit that it immediately began to build a million-dollar exhibit area and gift shop. The pandas were scheduled to arrive in June 1993.

San Diego wanted wild-caught pandas, despite the fact that AZA and FWS loan guidelines required the loan only of captive-born animals to prevent the capture of additional pandas to satisfy zoo demand. China complied with San Diego's unusual request. The Chengdu Zoo, one of the Chinese institutions involved in the loan, wrote to the San Diego Zoo in a letter dated 16 January 1992 that "as far as the rescue of wild pandas, it will be the zoo's major responsibility." "Rescue" has long been a euphemism for "capture." Two months after this letter, two wild pandas were "rescued," one of them apparently ill, and designated for the San Diego Zoo. By further coincidence, the "rescued" animals were a male and a female.

For too long, the panda rental program had suffered from noticeable apathy, as most people merely ignored or only mildly objected to exploiting the animals. But San Diego's commercialism created a furor in the conservation community. Other zoos also began to scuffle for pandas. Busch Gardens in Tampa made a preliminary agreement with China for a breeding pair, a ten-year loan for thirteen million dollars; however, that zoo was at least willing to delay importation for a while. And in May 1993, the Adventure World Zoo in Japan negotiated a ten-year, ten-million-dollar rental for a breeding pair.

Zoos commonly exchange endangered species on long-term breeding loans to enhance reproduction. If such exchanges take place as part of a coordinated program, a species can benefit greatly. On 19 January 1993, I attended a meeting of the AZA's Giant Panda Action Group, whose aim was to develop just such a program for the panda. The guiding principle of any plan would be "the long-term survival of giant pandas in nature." It was agreed that there had to be full integration of conservation efforts for pandas in the wild and in captivity as well as close cooperation between China and institutions elsewhere.

A breeding program could, of course, be conducted wholly in China, with other countries providing knowledge and equipment as needed. After all, China has a surfeit of breeding centers that now house over one hundred pandas. Once haphazard, China's breeding program has improved greatly in the past few years. In 1992, the most successful year to date, thirteen cubs were born and eleven survived. But 1993 produced only three births. The Wolong breeding center, after years of virtual failure, had six births between 1991 and 1993, and three cubs survived. Zoos also had success in hand-rearing the second young of a litter, which normally dies of maternal neglect. Wolong, Beijing, and Chengdu nurtured such cubs through the first few critical months. If such successes continue, the captive population could soon be self-sustaining, providing a surplus of pandas that could be used in loan programs and should be used to recolonize the wild. Improvement in reproduction is urgent because many of today's successfully breeding pandas are reaching old age.

In 1989, the World Wildlife Fund and the Ministry of Forestry published a panda conservation strategy. After various vicissitudes, the final version of the plan emerged from the Chinese government in April 1993. China allocated thirteen million dollars to the ten-year plan and sought an additional sixty-four million dollars from outside sources. The funds will be used to establish fourteen new panda reserves, move timber concessions, and provide financial incentives for resettling villagers and for other conservation programs. In my opinion the main justification for breeding loans is to provide funds for this conservation effort. Six panda pairs on ten-year loans could raise all the requested millions. And if the annual rental fee for a pair were two million dollars rather than one million, still a modest sum given the fact that pandas can represent commercial gold to a zoo, then a mere three pairs would have to emigrate from China for a decade. One problem has been the perception that so far China has not spent its panda rental income wisely.

The AZA tried to address this issue in a panda conservation action plan that Sydney Butler, executive director of the AZA, presented at a press conference in Washington, D.C., on 16 April 1993. I attended. The AZA proposed hiring a full-time coordinator to integrate the efforts of zoos, conservation organizations, and governments. An important part of the coordinator's job would be to help develop and monitor joint conservation projects with China to make certain that pandas actually benefit from the rental funds. As Devra Kleiman of the National Zoo told reporter Lena Sun of the *Washington Post:* "The North American zoo community is unlikely to turn over any money until we're comfortable that there's some mechanism for a unified decision-making process, accountability for funds and scientific management of the captive population." And, one should add, accountability for the wild population. Furthermore, the AZA expected to approve and coordinate all panda initiatives by North American zoos. The plan represented a useful first step, and a coordinator was appointed soon afterward.

But a truly workable survival plan depends also on the World Conservation Union, WWF, FWS, the Convention on International Trade in Endangered Species (CITES), and various Chinese institutions, all of which have their own panda policies and agendas. The AZA plan also had a provision that puzzled me: it endorsed the importation of wild-caught pandas. This placed it at odds with CITES and FWS regulations and reversed its own policy, as if AZA were accommodating the San Diego Zoo and endorsing its actions.

San Diego had expected its two pandas in June 1993. The month came and went, but the expensive panda exhibition area remained empty. The zoo's permit application was vigorously opposed by WWF and others on the grounds that the animals were wild-caught, the loan was not part of an integrated program, and there was no accountability for funds. The Reagan-Bush years of environmental neglect were over. For the first time in a dozen years the country had a secretary of the interior, Bruce Babbitt, who was familiar with and concerned about conservation issues. Political pressure was not now enough to obtain a panda permit. The FWS evaluated San Diego's application on its merit and on 21 September 1993 rejected it. "Denial of import permit threatens species" headlined the *San Diego Union-Tribune,* a newspaper whose panda articles all too often read as if written by the zoo's public relations department. Douglas Myers, director of the zoo, lamented in an interview that "if this decision sticks, truly the panda does not have a chance." And Jeff Jouett, a zoo spokesman, said: "We are not giving up on pandas. We are fighting back in every conceivable way." And with bizarre obstinacy, the zoo did. One of the pandas that had been designated for the zoo developed a fist-sized growth on its shoulder. San Diego asked the FWS for an emergency permit to bring the animal all the way to California for medical treatment. The FWS responded on 20 December by announcing that it would not process any more permit applications until it had developed its new panda policy. "The timing could not be better," AZA's Sydney Butler told a reporter.

If nothing else, this pitiful history of minor events raised public consciousness about the panda's plight. For example, the ABC television program *Day One* produced an excellent segment on panda rentals. When the president of San Francisco's Board of Supervisors, Angela Alioto, asked China for a panda loan, the *San Francisco Examiner* advised her to "stick to stuffed pandas. . . . Conservation, yes. Show biz (and politics), no."

On 20 November 1993 the AZA panda conservation committee produced a new version of its action plan, one designed to raise thirty to fifty million dollars for China. In part the plan states, "We will seek to bring up to a total of 17 animals to the U.S. on long term breeding loans, eight of them captive-born and nine wild-born/captive. . . . Breeding will focus the resources and experience of AZA member institutions on animals that have not bred successfully in China. . . . We also must ensure that income-generating loans do not stimulate further captures of pandas from the wild. Therefore, no animals taken into captivity after March 1993 will be considered for inclusion." San Diego and Busch Gardens received special consideration by the AZA in that they will be the first of the interested zoos to receive pandas under the new program.

If North America drains up to seventeen adult pandas from China, and other countries receive pairs as well, will China be able to maintain a viable breeding program of its own? I worry that more pandas will ultimately be abducted from the wild. The rental of only a few pairs could raise the millions of dollars sought by China, as I noted earlier. And if zoos donated all panda profits to conservation, the funds could benefit many endangered species and their habitats in China, not just the pandas. It is worth stating once again that panda loans, no matter how lofty the conservation rhetoric, remain commercial ventures in direct contravention of CITES and FWS regulations if a zoo makes money from the transaction. As Lisa Stevens of the National Zoo said to Associated Press reporter Brigitte Greenberg, "I think the interests of individual zoos need to be set aside . . . for the benefit of the species as a whole."

These events are only a few of the convolutions in American panda politics during the past two years, a period of endless memoranda and meetings, of progress without seeming change. But amidst the rivalry, greed, and compromises, dedicated individuals are shaping an acceptable panda policy.

Meanwhile China has made notable advances in panda conservation. Chinese officials may at first seem aloof and reluctant when pressured by outside ideas, but I have found that they always consider and, when possible, assimilate these ideas, especially if pragmatic suggestions conform to their concept of moral rightness. Initial progress in panda conservation was so slow that I was filled with creeping despair, but several recent actions by the Chinese government provide a margin of hope for the panda's future. Of fourteen planned new panda reserves, several have already been established and the rest soon will be. Ultimately about 95 percent of the pandas will reside in protected areas. Wolong now has a staff of forty rangers for anti-poaching patrols. Such a program must be extended to all panda forests to halt the lucrative trade in pelts. An international studbook was authored by Zhao Qingguo, Fan Zhiyong, Jonathan Gipps, and Devra Kleiman and published in 1992. Listing capture locations, birth and death dates, reproductive histories, and other pertinent information about specific individual pandas, it is an invaluable guide for improving management of the captive population.

In September 1993, Chengdu held an International Panda Festival, a combination of conservation conference and trade fair. By chance I passed through Chengdu at that time en route to Tibet. Flags with panda emblems festooned the streets and banners carried such slogans as "Five billion people love the 1,000 giant pandas sharing the earth." I met old friends like Bi Fengzhou, Pan Wenshi, and Lü Zhi but did not attend the conference. About a hundred participants from eleven countries came, including delegates from the American and Japanese zoo associations. The joint hosts were the China Association of Zoological Gardens and the China Wildlife Conservation Association, two organizations that usually find it as difficult to cooperate as some American zoos. All aspects of panda conservation were discussed at the conference. Afterward, twenty-nine participants from several nations sent a letter to Premier Li Peng. The letter noted that "within the last ten years the panda habitat has dropped by 50%. . . . The most critical threat to the survival of the natural population today is the timber harvest within habitat occupied by giant pandas. . . . We would like to learn that timber harvest will cease or be drastically reduced within panda reserves as soon as possible. We appreciate that there may be conflict between conservation and individual interests (as there is in developed and developing countries outside China), but we believe that the Chinese government would serve as a good example to the entire world in this important visible move." In November, the Chinese government responded favorably to the letter, instructing the Ministry of Forestry to halt timbering in critical panda areas and provide assistance in resettling people from such areas within the next three to four years. Some of the logging was indeed halted within a few months.

Fan Zhiyong of the Ministry of Forestry expressed the spirit of the conference when he stated, "Let us save our giant panda together."

In December 1992, for the first time in nearly eight years, I walked through a panda forest. Pan Wenshi and Lü Zhi had invited me to their study area in Gansu's Qinling Mountains where they have studied pandas since 1985. Their base was in the Chang Qing Forest Reserve, seventy-five square miles of rugged mountains reminiscent of our Tangjiahe research site. The morning of 11 December was crisp and clear as we traced the banks of a stream up a narrow valley. Bamboo and thickets covered the logged hillsides, but higher up, beyond dark cliffs, were ridges covered with oak, hemlock, and pine. Somewhere ahead was Jiao-Jiao, Double Beauty, a radio collared female who had given birth on 13 or 14 August. Jiao-Jiao had taken her cub from the birth den on 13 November and now was caching it here and there among the ravines. Lü Zhi had just returned from the United States, where she had been studying panda DNA in Stephen O'Brien's laboratory at the National Cancer Institute to determine genetic diversity in the species. She had not seen the cub since September and eagerly raced up the path.

Xiang Dingqian, a local project assistant, had left earlier to radio-locate Jiao-Jiao. After an hour's hike we came upon him waiting for us. The pandas were just ahead in a dense bamboo thicket. As we quietly approached, Lü Zhi whispered to me: "If she attacks, just call 'Jiao-Jiao' to her." Remembering Zhen-Zhen's displeasure on finding me near her cub in Wolong, I glanced around for a nearby tree. But Jiao-Jiao had dissolved into the bamboo, her radio signal the only evidence that she prowled silently nearby.

The cub was in a bower of bamboo at the base of a tree. We squatted down about five feet from her. Four months of age and as large as a chunky cat, she stood on bowed legs, displeased by our intrusion. She gave a screaming bark and aggressively swiped the air with a paw, then emitted a few baby whines and mews. Suddenly, as if dismissing us from her mind, she curled on her side, eyes closed, seemingly asleep. From fury to sleep in seconds. Such behavior puzzled me, but I did not want to destroy the harmony of her actions by trying to dissect the behavior.

Year after year
On the panda's face,
A panda's mask.

This paraphrase of Matsuo Basho's haiku about monkeys conveys the panda's enduring mystery. It was the first time ever that I could observe at leisure a cub in the wild. We sat there a long time just looking at this magic fluff of fur breathing calmly in her haven of bamboo.

Later, as we descended the trail, still entranced by the lustrous cub, we came upon Huzi, Little Tiger. Huzi was a male panda just over three years old, and a son of Jiao-Jiao. Though leading an independent life, he was still in the same area as his mother. Near him, in the low branches of a tree, was yet another panda, probably a female. Neither Pan Wenshi nor Lü Zhi knew her personally. Both pandas were filthy, as if they had wallowed in dirt, and Xiang Dingqian told us that he had seen them fight earlier that day.

It has taken Pan Wenshi and Lü Zhi years to know their pandas so well individually that they can now unravel the intricacies of their society as generation follows generation. I greatly admire their commitment to science and conservation. And I take a quiet pride in their research. We all strive for a private sense of merit, perhaps to do a study well or help protect a species. But the greatest satisfaction comes from leaving a country with an intellectual inheritance, imbuing a local biologist with a passion to continue the work. Pan Wenshi and I began panda research together in 1980, and we shared ideas and techniques for several years before he initiated his own study. As professor of biology at Beijing University, he has supervised a number of students, among them Lü Zhi, an extraordinarily capable and determined field researcher. She completed her doctorate on pandas in 1992 at the age of twenty-seven. Now on the staff of Beijing University, she will in turn teach students, the knowledge and spirit of our original collaboration flowing onward and outward long after my fleeting presence in China is at most a cloud shadow of memory.

As I write this in April 1994, Jiao-Jiao still roams the Qinling Mountains with her cub, her gift to the future. The cub's name is Xiwang, and the name means Hope.

Selected Reading

Books

The following books provide useful background information about pandas and the area in which they live.

Catton, Chris. 1990. *Pandas*. New York: Facts on File. A good overview of pandas by an author who did his homework both in the field and the scientific literature.

Harkness, Ruth. 1938. *The Lady and the Panda*. London: Nicholson and Watson. The story of Su-Lin, the first panda to reach the Western world.

Fox, Helen, ed. 1949. *Abbé David's Diary*. Cambridge: Harvard University Press. The panda discoverer's diary, 1866–69, edited and translated from the French.

MacClintock, Dorcas. 1988. *Red Pandas, a Natural History*. New York: Scriber's. A fine summary of knowledge about this attractive and interesting species.

Morris, Ramona, and Desmond Morris. 1966. *Men and Pandas*. New York: McGraw-Hill. An excellent summary of panda biology and of Western involvement with the panda from 1869 to the 1960s. (The book was revised by J. Barzdo and reissued as *The Giant Panda* in 1981.)

Roosevelt, Theodore, and Kermit Roosevelt. 1929. *Trailing the Giant Panda*. New York: Scriber's. An account of the first successful panda hunt by Westerners.

Schaller, George, Hu Jinchu, Pan Wenshi, and Zhu Jing. 1985. *The Giant Pandas of Wolong*. Chicago: University of Chicago Press. Research results that cover the first years of the panda project.

Sheldon, William. 1975. *The Wilderness Home of the Giant Panda*. Amherst: University of Massachusetts Press. An account of the 1934 Sage Expedition which shot a panda.

Smil, Vaclav. 1984. *The Bad Earth*. Armonk, N.Y.: M. E. Sharpe. A record of environmental degradation in China.

Tang Xiyang. 1987. *Living Treasures: An Odyssey through China's Extraordinary Nature Reserves*. New York: Bantam Books. Essays about several of China's most beautiful nature reserves, including Wolong, with many color photographs.

Wilson, E. H. 1986. *A Naturalist in Western China*. London: Cadogan Books. This reprint of the 1913 edition is a key source of information about travel, people, and natural history in panda country during the early part of this century.

Articles

The panda project produced many popular and scientific articles which complement this book with information and photographs. The list below presents a selection of articles published between 1981 and 1991.

Popular articles

Johnson, Kenneth, and Jenny F. Johnson. 1990. "Mystery of the Other Panda." *International Wildlife*. 20(6):30–33.

Machlis, Gary, and Kenneth Johnson. 1987. "China: Panda Outposts." *National Parks*. 61(9–10):14–16.

Reid, Donald. 1988. "Some Pandas Are Red." *Dinny's Digest* (Calgary Zoo). Spring. Pp. 3–11.

Schaller, George. 1981. "Pandas in the Wild." *National Geographic*. 160(6):735–49.

———. 1982. "Zhen-Zhen, Rare Treasure of Sichuan." *Animal Kingdom*. 85(6):5–14.

———. 1985. "China's Golden Treasure [golden monkeys]." *International Wildlife*. 15(1):29–31.

———. 1986. "A Face Like a Bee-stung Moose" [Takin]. *International Wildlife*. 16(6):36–39.

———. 1986. "Secrets of the Wild Panda." *National Geographic*. 169(3):284–309.

———. 1987. "Notes of a Professional Panda-Watcher." *Animal Kingdom*. 90(4):20–25.

———. 1988. "Towards the Dark Abyss of Extinction." *World Magazine* (England). July. Pp. 23–28.

Scientific articles

Campbell, Julian, and Qin Zisheng. 1983. Interaction of giant pandas, bamboos, and people. *Journal of the American Bamboo Society*. 4:1–35.

De Wolf, Robert, Roland Goossens, John MacKinnon, and Wu Shen Cai. 1988. Remote sensing for wildlife management: giant panda habitat mapping from LANDSAT MSS IMAGES. *Geocarto International*. 1:41–49.

Johnson, Kenneth, George Schaller, and Hu Jinchu. 1988. Responses of giant pandas to a bamboo die-off. *National Geographic Research*. 4:161–77.

———. 1988. Comparative behavior of red and giant pandas in the Wolong Reserve, China. *Journal of Mammalogy*. 69:552–64.

Reid, Donald, Hu Jinchu, Dong Sai, Wang Wei, and Huang Yan. 1989. Giant panda *Ailuropoda melanoleuca* behavior and carrying capacity following a bamboo die-off. *Biological Conservation*. 49:85–104.

Reid, Donald and Hu Jinchu. 1991. Giant panda selection between *Bashania fangiana* bamboo habitats in Wolong Reserve, Sichuan, China. *Journal of Applied Ecology*. 28:228–43.

Reid, Donald, Hu Jinchu, and Huang Yan. 1991. Ecology of the red panda *Ailurus fulgens* in the Wolong Reserve, China. *J. Zoology*, London. 225:347–64.

Reid, Donald, Jiang Mindao, Teng Qitao, Qin Zisheng, and Hu Jinchu. 1991. Ecology of Asiatic black bears (*Ursus thibetanus*) in Sichuan, China. *Mammalia.* 55:221–37.

Reid, Donald, Alan Taylor, Hu Jinchu, and Qin Zisheng. 1991. Environmental influences on bamboo *Bashania fangiana* growth and implications for giant panda conservation. *Journal of Applied Ecology.* 28:855–68.

Schaller, George, Teng Qitao, Pan Wenshi, Qin Zisheng, Wang Xiaoming, Hu Jinchu, and Shen Heming. 1986. Feeding behavior of Sichuan takin (*Budorcas taxicolor*). *Mammalia.* 50(3):311–22.

Schaller, George, Teng Qitao, Kenneth Johnson, Wang Xiaoming, Shen Heming, and Hu Jinchu. 1989. The feeding ecology of giant panda and Asiatic black bear in the Tangjiahe Reserve, China. Pp. 212–41. In *Carnivore behavior, ecology, and evolution,* ed. John Gittleman. Ithaca: Cornell University Press.

Taylor, Alan, and Qin Zisheng. 1988. Regeneration patterns in old-growth *Abies-Betula* forests in the Wolong Natural Reserve, Sichuan, China. *Journal of Ecology.* 76:1204–18.

———. 1988. Tree replacement patterns in subalpine *Abies-Betula* forests, Wolong Natural Reserve, China. *Vegetatio.* 78:141–49.

———. 1988. Regeneration from seed of *Sinarundinaria fangiana,* a bamboo, in the Wolong Giant Panda Reserve, Sichuan, China. *American Journal of Botany.* 75:1065–73.

———. 1989. Structure and composition of selectively cut and uncut *Abies-Tsuga* forest in Wolong Natural Reserve and implications for panda conservation. *Biological Conservation.* 47:83–108.

Taylor, Alan, Donald Reid, Qin Zisheng, and Hu Jinchu. 1991. Spatial patterns and environmental associates of bamboo (*Bashania fangiana* Yi) after mass-flowering in southwestern China. *Bulletin of the Torrey Botanical Club.* 118:247–54.

Taylor, Alan, Donald Reid, and Qin Zisheng. 1991. Bamboo dieback: an opportunity to restore panda habitat. *Environmental Conservation.* 18:166–68.

Index of Names

Index of Species and Subjects

The Wildlife Conservation Society has a global mission to preserve biological diversity. Its biologists conduct over one hundred and twenty-five projects in more than forty countries to develop innovative approaches to conservation. The Wildlife Conservation Society integrates the needs of wildlife and people, trains local biologists, devises plans for protecting wildlife and critical habitats, and studies key species, such as the giant panda.

You can help the Wildlife Conservation Society save wild animals and wild lands around the world. For more information, contact:

Wildlife Conservation Society
Bronx Park, Bronx, NY 10460
(718) 220-6891